Technical Math

FOR

DUMMIES®

Technical Math
FOR
DUMMIES®

by Barry Schoenborn and Bradley Simkins

WILEY

Wiley Publishing, Inc.

Technical Math For Dummies®

Published by
Wiley Publishing, Inc.
111 River St.
Hoboken, NJ 07030-5774
www.wiley.com

For general information on our other products and services, please contact our Customer Care Department within the U.S. at 877-762-2974, outside the U.S. at 317-572-3993, or fax 317-572-4002.

For technical support, please visit www.wiley.com/techsupport.

Wiley also publishes its books in a variety of electronic formats. Some content that appears in print may not be available in electronic books.

Library of Congress Control Number: 2010926845

ISBN: 978-0-470-59874-0

Manufactured in the United States of America

10 9 8 7 6 5 4 3 2 1

WILEY

About the Authors

Barry Schoenborn lives in Nevada City, California. He's a longtime technical writer with over 30 years' experience. He's written hundreds of user manuals and (in the early days) worked dozens of part-time jobs that required practical math. He has been a carpenter for the movies, a stage electrician, a movie theater manager, a shipping clerk, an insurance clerk, and a library clerk. He has a bachelor's degree in theatre from California State University, Fullerton.

Recently, his company worked with the California Integrated Waste Management Board to teach scientists and administrators how to write clearly. Barry is the coauthor of *Storage Area Networks: Designing and Implementing a Mass Storage System* (Pearson Education). He was a movie reviewer for the *L.A. Herald-Dispatch* and wrote a monthly political newspaper column for *The Union* of Grass Valley, California, for seven years. Barry's publishing company, Willow Valley Press, published *Dandelion Through the Crack*, which won the William Saroyan International Prize for Writing.

Bradley Simkins was born and raised in Sacramento, California, and became a sixth-generation journeyman plasterer. But it didn't take long (after many hours on construction sites) before he figured out that it was easier to use his brain than his muscles. He has a master's degree in mathematics from California State University, Sacramento. He has taught, assisted, and tutored at the Multimedia Math Learning Center at American River College in Sacramento. He and his family live in Sacramento, where he owns Book Lovers Bookstore, an independent bookstore.

Dedications

Barry: To my teachers at San Juan High School in Citrus Heights, California: Mr. N. E. (Norm) Andersen (math); Mrs. Eada Silverthorne (English); Ms. Susan A. Schwarz (English); Mr. Norman E. Allen (physics); Mr. A. J. Crossfield (chemistry); and Mr. James C. Harvey (biology). They would be surprised and (maybe) pleased.

Bradley: I dedicate my work to my in-laws, Greg and Diane Manolis, who have always extended their hand to help with no complaints, and to my oldest daughter, Ashleigh, who taught me that failing does not make you a failure.

Authors' Acknowledgments

Barry: This book wouldn't have been possible without the efforts of coauthor Bradley Simkins. We were supported by a great team at Wiley Publishing (Natalie Harris, Erin Mooney, and Megan Knoll) who worked hard to make this book a reality. They are the nicest people you'll ever meet! A big thanks, too, to Matt Wagner of Fresh Books Literary Agency, who presented us to Wiley.

Our patient readers were Priscilla Borquez (who is fast, accurate, and sensible, and who also has a great sense of humor); Jim Collins (an excellent and thorough technical communicator); Bill Love (who knows a zillion things about cars, machining, and welding); and Frances Kakugawa (author, poetess, and lecturer to and supporter of Alzheimer's caregivers, who was our poster child for someone who doesn't understand story problems).

Many thanks to Patricia Hartman, who was always encouraging, and to Johna Orzalli, my haircutter, who taught me how to mix hair color. Thanks as well to Jeff Perilman at Dave's Auto Repair for tips about smogging a car and specialized tools. And, finally, thanks and apologies to all the medical and dental staffs I flooded with questions.

Bradley: First, I thank Barry Schoenborn for all his hard work and dedication to make this work possible and for always going the extra mile to understand when my life became too hectic. I thank my beautiful wife, Audrey, and my beautiful children, Ashleigh, Brayden, Alexander, and Natalie, who make my life worth living. Thanks to Jill Marcai and Jens Lorenz for correcting all of our math mistakes. Last, but certainly not least, many thanks to the team at Wiley Publishing for taking on such goofballs.

Publisher's Acknowledgments

We're proud of this book; please send us your comments at http://dummies.custhelp.com. For other comments, please contact our Customer Care Department within the U.S. at 877-762-2974, outside the U.S. at 317-572-3993, or fax 317-572-4002.

Some of the people who helped bring this book to market include the following:

Acquisitions, Editorial, and Media Development

Project Editor: Natalie Harris

Acquisitions Editor: Mike Baker

Copy Editors: Megan Knoll, Caitie Copple

Assistant Editor: Erin Calligan Mooney

Editorial Program Coordinator: Joe Niesen

Technical Editors: Jens Lorenz, Jill Macari

Editorial Manager: Christine Meloy Beck

Senior Editorial Assistant: David Lutton

Editorial Assistants: Rachelle Amick, Jennette ElNaggar

Art Coordinator: Alicia B. South

Cover Photos: Corbis

Cartoons: Rich Tennant (www.the5thwave.com)

Composition Services

Project Coordinator: Patrick Redmond

Layout and Graphics: Carrie A. Cesavice

Proofreader: Jennifer Theriot

Indexer: BIM Indexing & Proofreading Services

Publishing and Editorial for Consumer Dummies

Diane Graves Steele, Vice President and Publisher, Consumer Dummies

Kristin Ferguson-Wagstaffe, Product Development Director, Consumer Dummies

Ensley Eikenburg, Associate Publisher, Travel

Kelly Regan, Editorial Director, Travel

Publishing for Technology Dummies

Andy Cummings, Vice President and Publisher, Dummies Technology/General User

Composition Services

Debbie Stailey, Director of Composition Services

Contents at a Glance

Table of Contents

Introduction

· ·

*T*echnical careers require technical mathematics (technical math). That's why we wrote *Technical Math For Dummies*. Whether you're currently working in a technical trade or studying in school, you have probably made the discovery that most jobs require some math.

Most parts of technical math are simple. You may think some parts are hard, but look closer. After you read them, you'll hit your forehead with the heel of your palm and say, "Yes! Of course! I sorta knew that all along, but now I really get it!" We think we've filled a gap in the world of math guides, and we hope you enjoy the book.

About This Book

This book is a reference. It's also a repair manual that can help you fill voids you may have in your math background. It's different from other math books in three major ways:

- ✔ **It's all about practical math.** You won't find anything about symplectic geometry or sigma-algebra here. Our focus is on math for technical careers — it looks at problems you may deal with every day and the math skills you need to handle them. But we also include general principles when necessary.

- ✔ **It's comprehensive.** It covers all major math concepts; other math books are about individual concepts (for example, algebra, geometry or trigonometry).

- ✔ **It's not dull** (we hope) as other math books often are. One of us (Barry) is a long-time technical writer, and he's written far too many deadly dull user manuals. That nonsense stops here. Because it's a *For Dummies* book, you can be sure it's easy to read and has touches of humor.

Technical Math For Dummies applies basic math to basic tasks in many careers. You get practical examples, and most of them are based on real-life experiences. And in what other book can you work with math and also find out how to make 90 dozen pralines or figure the distance from a fire watch tower to a wildfire? You can also apply a lot of this math to your personal life as well as your work life.

At the risk of sounding like a late-night infomercial, we want to point out a couple of this book's unique features. We gar-on-tee you won't find them anywhere else.

Conventions Used in This Book

We designed this book to be user-friendly, maybe even user-affectionate. If it were any friendlier, it would drive itself to your house and bring coffee and doughnuts. To help you get the most out of your new friend, we use the following conventions:

- *Italic type* highlights new terms. We follow each term with a short and often informal definition. Occasionally, we give you clues about how to pronounce difficult words.

- Web addresses are in `monofont`. They're usually very short and shouldn't break across two lines of text. But if they do, we haven't added any extra characters (such as a hyphen) to indicate the break. Just type in what you see.

- Although our English teachers would cringe at our breaking the rules, we usually write numbers as numerals, not words. For example, the text may say "add 9 to 3 to get 12," not "add nine to three to get twelve." We think this setup makes the ideas clearer in a math book.

What You're Not to Read

We'd love for you to read every word in this book in the order it appears, but life is short. You don't have to read chapters that don't interest you. This reference book is designed to let you read only the parts you need.

You don't have to read anything with a Technical Stuff or Did You Know? icon. That text is there to give you overly technical or trivial info. *Sidebars* (that's what they're called in publishing) are the shaded blocks of text you find every so often throughout the book. They're interesting (we think) but not critical to your understanding of the main text, so you can skip 'em if you want.

Foolish Assumptions

Although we know what happens when you assume, we went ahead and made a couple of presumptions about you anyway:

- ✔ We assume that you went to elementary and middle school, where you were exposed to math fundamentals. Why don't we include high school? Because high school is where many people get bored, dazed, or frustrated with mathematics. You may have been in class, but maybe your mind was somewhere else.

- ✔ We assume you have access to a computer and the Internet. It's not essential, but it's very handy. Use a good search engine to find out more about any topic in this book.

How This Book Is Organized

Technical Math For Dummies has five parts, moving from simpler topics (such as counting) to more complex topics (such as trigonometry). Here's how it's set up.

Part 1: Basic Math, Basic Tools

In this part, you get math basics (and we do mean basics). Chapter 1 gives you an overview of broad technical math concepts. Chapter 2 dispels myths about math and provides some history about technical careers. Technical professions are very old and go back (at least) to making arrowheads and spear points. And with all due respect to art history and library science majors, stonemasons built the pyramids. You also learn about the tools of the trades in this chapter.

The remaining chapters in this part are a complete review of basics — numbers, addition, subtraction, multiplication, division, measurement, and conversion. You see how to do these operations faster and better. We also tackle something that everybody says fills them with fear and loathing — the notorious word or story problem. Story problems can be filled with tricks and traps, but in this chapter you see how easily you can deal with them all.

Part II: Making Non-Basic Math Simple and Easy

In Part II, you review the workhorses of technical math, the processes that are a simple step above arithmetic. Most careers can't function without them.

Part III: Basic Algebra, Geometry, and Trigonometry

You may think some topics (algebra, geometry, and trigonometry) are tough, but in this part you find out that they aren't. The basic techniques are easy to understand, and those are the techniques you need. Now that's a happy coincidence!

This part removes the mystery from formulas and shows you how to make your own custom formulas. It's also filled with practical applications for areas, perimeters, and volume, as well as a little theory.

Part IV: Math for the Business of Your Work

In Part IV, we point out the obvious: "Life math" is different from "classroom math." Although the previous parts have direct application to your technical work, this part brings some math concepts to the business side of your job.

In this part, you see how to use graphs and charts to your advantage for both problem solving and presenting information to management and clients. We also present a chapter on time math, which we hope clears up a few mysteries about the basic questions "What time is it?" and "How long will it take?"

The last chapter deals with computer math, and it's a simple mini-education in what's going on with your computer and your Internet connection. This chapter may help make you a smarter shopper when you're buying computers, smartphones, MP3 players, and digital instruments for your business or your home.

Part V: The Part of Tens

For Dummies books always have a Part of Tens, and this book is no exception. The world loves lists of ten things, and in these chapters you find a large amount of information in a small space.

Chapter 20 has ten principles for solving any common math problem. Its partner is Chapter 21, which contains the ten most commonly used formulas. It also has some formula variations and some estimating shortcuts. Finally, Chapter 22 shows you ten easy ways to get good at math while doing everyday tasks. Finally, we also include a glossary of terms that you may or may not see in the text but that may pop up in your work.

Icons Used in This Book

We use several *icons* (the little drawings in the margins of the book) to call out special kinds of information and enhance your reading experience — that's just the kind of people we are. Here's a breakdown:

A Tip is a suggestion or a recommendation that usually points out a quick and easy way to get things done.

This icon represents a key idea that's worth remembering — the information may come in handy later.

Technical Stuff contains information that's interesting but overly technical and not vital to your understanding the topic.

Text with this icon contains odd facts (such as a legislature trying to regulate the value of pi), pieces of pop culture, strange bits of history, or bizarre terms.

The text with this icon describes a situation where a math principle is used in real-world work.

This icon alerts you to conditions that can spoil your work or result in wrong answers. For example, dividing by zero is never allowed in math. Don't try it or your hair may catch fire!

Where to Go from Here

You can go to any chapter of the book from here. First, check the table of contents, where you see the names of the parts and the chapters. Then, pick a chapter you're interested in.

The book isn't linear, so you can start anywhere. If you're comfortable with some math concepts, take a glance at the early chapters of this book. This strategy will confirm how much you already know (and you may pick up a couple of interesting new words, too). Then go on.

If you're uncomfortable with some math concepts (and some of them have truly bizarre and intimidating names), take a look at those chapters. Inside every "complicated" math concept is a simple concept trying to get out.

If you get stuck, you'll probably find another chapter that can help you out. If you haven't made a choice, we recommend beginning with Chapter 1, which introduces the broad concepts. If you have a particular problem, find a chapter in the table of contents that deals with it and go straight to it, or simply look up that topic in the index.

Part I
Basic Math, Basic Tools

The 5th Wave   By Rich Tennant

"Ed's in charge of calculating your height and weight ratio so you get the right length Bungee cord. He's too stupid to do anything else."

In this part . . .

Part I starts with the basics. In Chapter 1, you find the broad scope of what technical math involves. Chapter 2 identifies the myths of math and the trades that make the world as you know it possible from earliest to latest. It also gives you a survey of tools (especially new digital tools) that make your work (particularly measuring) in the trades easier and more fun.

The other chapters in this part offer a complete review of numbers and arithmetic. But they're more than just a good review — they also give you new insights and may even speed up your work. Chapter 6 is about measurement and conversion. Sorry to say it, but the world speaks measurement in different units, and the modern technician needs to know unit conversions. Chapter 7 is about word problems. After you read this chapter, you'll never run from a story problem again.

Chapter 1

Math that Works
as Hard as You Do

Technical mathematics (technical math) is an essential part of the work and the education of everyone in a technical career. If you're studying a trade in a two-year college or an occupational program, you can't dodge it, whether you're taking formal math courses or dealing with math calculations in specialized courses.

For example, Heald College is a famous college in San Francisco and much of California in general, as well as Portland and Honolulu. To get a degree as a medical assistant (Associate in Applied Science), you have to take Math 10, Essential Math; Math 103, Elementary Algebra; and Math 205, Modern Business Mathematics. And that's for an education in healthcare.

Even if you're already working in the field you want, you encounter plenty of technical math to do. All the construction trades deal with math to build buildings, pour sidewalks, install flooring, lay carpet, calculate fencing runs, and figure out how much paint goes on the walls. And because these trades are businesses, you have to figure amounts of materials, costs of materials and labor, and client billing. You may try to avoid math, but if you do, you may be avoiding a chance to advance your career.

Bottom line: Math isn't just something theoretical that professors in universities work with. It's a practical skill used in most careers. Technical occupations built the world and also made it fit to live in. That takes technical math.

If math gives you the willies, heebie-jeebies, butterflies in the stomach, or palpitations of the heart, suffer no more. Every principle in this book is easy, if you look at it the right way. Technical math is easier (not harder) than you think.

Discovering the Benefits of a Technical Math Book

Regular math books are fine, but we believe that you can get more benefit more quickly from a technical math book. A technical math book is all about practical math, focusing on math for technical careers — the math principles you're likely to need in everyday work. Abstract math need not apply.

Unlike regular math books, which tend be about a single discipline (for example, algebra, geometry, or trigonometry), a technical math book is comprehensive so that you don't have to go to several texts to get what you need. This book covers a little bit about a lot of subjects, and no subject goes deeper than you need it to go.

A good technical math book also includes practical examples based on real-life experiences. As a result, you may even discover something about careers other than your own. And you may be able to apply a lot of workplace math to your personal life as well as your work life.

The Basics Are Basically Basic

The most basic component of math is *numbers*. The first thing you do with numbers is count, and you started counting when you were very young — as soon as you could talk, your mother probably cajoled you to tell Aunt Lucy how old you were or to count from one to five.

If you put numbers on a line, you get (are you ready?) a *number line,* shown in Figure 1-1. The number line is an arrangement of whole numbers called *integers.* (See Chapter 3.) With a number line, you can count as high as you want by going to the right and as low as you want by going to the left.

Figure 1-1:
A number
line.

Counting is not only the first math thing you probably ever did, but it's also the first thing that ancient people did. The earliest math discovery is the Ishango bone, a tally stick, and it's more than 20,000 years old!

Another basic component of math is *arithmetic*. That's addition, subtraction, multiplication and division. You learned them in elementary school, but if you didn't understand them well, you may still have trouble with the processes today. Even if you knew them and then forgot them, you can get a refresher in Chapters 4 and 5.

The word *arithmetic* comes from the Latin word *arithmetica,* which comes from the Greek words for "counting," "number," and "art." Yes, it's the art of counting numbers.

When you know about numbers and know arithmetic, you're on your way to becoming a technical math terror. They're basic skills, but those basic skills handle a lot of the math in day-to-day life and prepare you for some more interesting topics.

Meeting Measurement and Conversions and Studying Story Problem Strategies

Measuring quantities and amounts is fundamental to every career; knowing your units is important. It sounds simple, but the world throws you a couple of curve balls with two different common systems of measurement (American and metric), which we cover in Chapter 6.

Chapter 6 also shows you all the basic units of length, area, weight, volume, and liquid volume and how to convert from one unit to another, an essential in technical work. You also see how to convert from one unit system to the other.

When you know your math basics and your conversions, you can slay math monsters faster than Conan the Barbarian slays movie monsters. But Conan had a vital tool you don't have, the Atlantean Sword. That's where Chapter 7 comes in. It contains the tricks, traps, and techniques you need for solving story problems; with its help, you'll laugh, scoff at, deride, and mock the so-called word problems that come up in everyday work life.

Using Workhorse Math

Four math disciplines — fractions, decimals, percentages, and exponents — are the workhorses of your trade. All careers use one or more of them, and some careers use all of them.

You use these four workhorses for many utilitarian purposes, a little like the way draft horses have been used for logging, plowing, pulling beer wagons (yes, the Budweiser Clydesdales are draft horses), hauling freight, and transporting passengers in horsecars. Like the horses, the math disciplines are strong and docile.

The chapters in Part II of this book tell you plenty about fractions, decimals, percentages, and exponents (and the exponent's trusty sidekick, the square root), but here's a little taste to whet your appetite:

- **Fractions:** Fractions come in various forms, including stacked, unstacked or inline, decimal, and percentage; check out the following for examples.

  $$\frac{1}{2} \quad \frac{1}{2} \quad 0.500 \quad 50\%$$

  You use fractions in just about every trade, and not only in doing your basic job — this math also comes up in working with time, money, and computer capacity. Find out more in Chapter 8.

- **Decimals:** Decimals are a form of fraction, and they're essential for work in major trades. The laboratory and the machine shop are two places where you find a lot of decimal numbers. Chapter 9 gives you the details.

- **Percentages:** Percentages are fractions based on 100. You need percentages to express portions of a whole quantity, and they're at the very core of working with money. Head to Chapter 10 for more.

- **Exponents and square roots:** Exponents let you express very big and very small numbers (and do math with them) in a very compact way. Square roots help you solve a couple of pesky problems in your daily work. Chapter 11 has the lowdown on these concepts.

Building Your Knowledge of the Branches of Math

Some people say, "A little knowledge is a dangerous thing," but that's not necessarily so. *No knowledge* is a dangerous thing.

Don't worry if you have "a little knowledge" about math. The good news is that you only need to know a little and you'll do fine. But perhaps the best news is that Part III helps you out by expanding what you do know.

Algebra makes any problem solvable after you figure out the formula you need. As Chapter 13 shows, formulas are easy to develop, and they make even complicated story problems (shown in Chapter 7) collapse into solutions.

Geometry, as shown in Chapter 14, lets you draw the various shapes you need to measure landscape jobs, dress patterns, or whatever your job requires. Plus, you develop a great vocabulary about lines, angles, and shapes that can aid you in your daily work.

Alexander Pope, the poet, first wrote this saying in *An Essay on Criticism* in 1709.

> A little learning is a dangerous thing;
> drink deep, or taste not the Pierian spring:
> there shallow draughts intoxicate the brain,
> and drinking largely sobers us again.

You may even use the Pythagorean theorem to find the lengths of sides of a triangle. That's part of a method for finding the areas of patios, yards, and odd-shaped rooms, and in special cases, you can even use it to find the area of a piece of pie. Figure 1-2 shows a classic geometrical view of the theorem.

Figure 1-2:
Graphic
of the
Pythagorean
theorem.

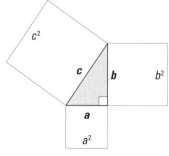

In Figure 1-2, a^2 represents the square of length a, and b^2 is the square of length b. Also, c^2 is the square of side c. Don't worry about the letters or even the theorem now — we reveal all in Chapter 14. And check out Chapter 15 for help with areas, perimeters, and volumes.

In trigonometry, good math appears to be magic, but it's really just good math. Trigonometry is essential for surveyors, land engineers, and fire lookouts, to name just a few. When you do a little trig, you can easily figure out how wide a river is without getting your feet wet. The solution to this problem has eluded one of the authors (Barry) since he was Boy Scout, but with the help of Chapter 16, he (and you) can finally cross that bridge.

Life Math Isn't Classroom Math

The math of the classroom is good. The principles are solid, and the math is conceptual as well as real. Classroom math improves your thinking, and improved thinking can greatly reduce the Homer Simpson "D'oh!" factor in your life.

However, the math of life is what you face every day. It's good, real, and entirely practical. When you do life math, it directly affects your work and the people who depend on you. Your calculations can affect

✔ The appearance and building quality of a client's new home

✔ Effective wildfire fighting

✔ Precise property line measurements

✔ Accurate reporting of patients' vital signs

✔ Correct dispensing of drugs to patients

✔ The quantity, taste, and nutrition of what people eat

Luckily, the chapters in Part IV help you deal with this side of math. They help you use graphics (particularly charts and graphs), do excellent time accounting (for payroll and client billing), and make smart purchases of computers and high-tech instruments (computer math).

"But wait," you say. "I don't do payroll or buy the computers, and no one has seen a graph at my office since 1972. Why do I need all this stuff?" The answer: Sometimes these math tools can be very valuable in your personal life. Pretend (and it's not really pretending) that your wages seem flat, yet your family's health insurance and out-of-pocket healthcare costs seem to be rising. Is that true? Find out by using a line graph (which we cover in Chapter 17). Figure 1-3 shows a comparison of annual wages and annual healthcare costs over several years.

The graph clearly shows something you'd hoped not to see. Healthcare costs are in fact overtaking your salary at a rapid rate.

Wages and Healthcare Costs

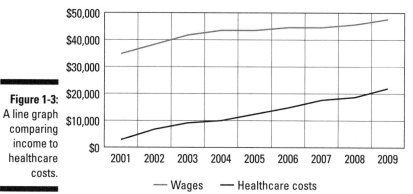

Figure 1-3: A line graph comparing income to healthcare costs.

Chapter 2

Discovering Technical Math and the Tools of the Trades

*M*athematics is useful and fun. Read that sentence again, because it may startle you. Math is useful and fun, and it can get you out of more trouble than Superman and Batman combined. The main reason math is useful is that you can do so much with it; it's a practical tool for solving problems in many careers. Finding answers to the questions and concerns that come up on the job is really satisfying. You get a great feeling when the light goes on and you say, "Oh! I get it!" And on the job, success is supposed to come to the person who gets the most things right. The trouble is, math gets a bad rap. No one knows who first started to give such a nice skill such a bad reputation — the search still goes on for whoever started spreading lies about math. As a result of this scoundrel, some urban legends about math still persist.

The technical work you do is vital in a world that relies on technology. And highly skilled technical work requires tools to get the work done. Look at the tools you use for technical math. Some are general, while others are very specific. Some simply measure, others calculate, and some do both. So what are the tools of the trades? They vary from career to career, but everyone uses a couple of tools. The two most important tools are your general math skills and the modern calculator. You need to know the math so you can appreciate what the specialty instruments do for you, and so you can do the same operations if you don't have such instruments handy.

In this chapter, you find out what technical math is all about. Add to that a little history, because if people have been doing this stuff for so long, how hard can it be? And you also get a view of what tools of the trades are available.

Busting Myths about Math

Many myths about math spring from two great myths. If you're able to recognize those myths, you're well on your way to busting math myths yourself. The following sections introduce you to "I don't need to use it" and "It's too hard," along with their cousin, "I have a phobia."

"I don't need to use it."

Don't mean to be rude, but talk to the hand, 'cause the face ain't listening. Yes, you do have to use math. First, you need math to get through most programs of education (trade-related or otherwise). Then you likely need math on your job; even if the math is limited to counting and measuring, it's math (and you can be sure we cover counting and measuring in this book).

You may think no math is involved in the culinary arts (the world of cooking, pastry, baking, and candy making), but think again. With all due respect for hamburger flippers, there's a world of difference between grilling a double bacon cheeseburger and being executive chef at a three-star restaurant. The difference comes from a completing culinary school, which requires (wait for it) math.

The Michelin Guide started awarding stars to the best restaurants in 1926, and stars aren't easy to get. The 2010 guide lists only 25 three-star restaurants in France and only 85 in the world.

Executive chefs do more than cook. They create, plan, budget, and do cost accounting. The work is a combination of art, cooking, and math. Even the food service operation of a hospital or retirement home requires math, such as scaling recipes up (as in turning a lasagna recipe that serves 6 into one that serves 300) and down (turning a recipe that serves 300 into one that serves 6).

The "I don't need to use it" myth makes no sense to carpenters, cabinetmakers, concrete masons, lab technicians, cooks, or surveyors, whose careers clearly require math to get the job done. The only exception may be those who don't want to advance in a career. If you don't want to go anywhere, don't study math.

To be much more positive, if you *do* want to go somewhere, *do* study math. And take heart! Some fields require only simple arithmetic (addition, subtraction, multiplication, and division), which you need for everyday life anyway.

"It's too hard."

Another grand math myth is "It's too hard." This myth was probably started by a person who said that about everything. Life isn't particularly easy, so the mantra probably got a big reception from everybody, not just those who were doing math. Survival isn't easy. If you go back a zillion years (that's an approximate date), you see that hunting for food was "too hard," yet somehow the human race hunted, survived, and got civilized.

You know what's hard? Walking is hard. A human being isn't constructed all that well for walking, but the average child learns how to walk by the time she's about 11 months old. She looks a little clumsy at first but later becomes very proficient at walking as she does more of it.

Driving a car is hard, but most people can do it. They learn how to drive, and then (get ready) they drive. Two things make you a good driver: knowing the fundamentals and practicing.

That concept, of course, is shared by every professional dancer and athlete in the world. And it applies to just about every action or operation a professional does. It's the same with mathematics. Know the fundamentals and practice. It's a mantra you can live by. The carpenter's first try at driving a nail probably bends the nail, and his first saw cut is probably crooked. But knowing the fundamentals and practicing eventually make the difference.

"I have a phobia."

This misconception is a variation of the popular statements of denial in the preceding sections. Unfortunately, the argument doesn't have legs, because all people approach new experiences with anxiety.

The distinction between anxiety and phobia is important. *Anxiety* is an everyday emotion. A *phobia* (a fear) is an anxiety disorder. Phobias are the most common form of anxiety disorders. In a study, the National Institute of Mental Health found that between 8.7 percent and 18.1 percent of Americans suffer from phobias.

You may have a phobia, but actual, legitimate *math phobia* (fear of not succeeding at math) isn't common, regardless of what Internet hucksters try to tell (and sell) you. To be fair, math phobia does exist, but it's not a permanent condition; after sufferers experience even small amounts of math success, they usually overcome it. So really, having a math phobia is even *more* reason to do math.

Remember: Somebody Else Already Did the Hard Work

Civilization makes math. But here's a paradox. Math makes civilization. Mathematics started a long time ago in a galaxy far away (well, actually, on this planet). In the beginning, math was just about counting. (See Chapter 3.) And for a time in human existence, that was all people needed.

Well, being a hunter-gatherer is all very fine, but (to tweak an old song), we can show you a better time. That "better time" was nice, stable agriculture, which required some basic math to make it work when it was established about 10,000 years ago. Farming settles people down. It starts cities growing and it also produces specialized trades.

Math lets a culture do more, have more, and be more. As cultures grew more civilized, they needed to measure land and trade with other cultures. That requires math, so technical math grew and got sophisticated.

In the mere thousands of years from the cave dwellers to texting, mathematicians made discovery after discovery. They not only figured concepts out but also did what are called rigorous mathematical proofs. To put it another way, if you read about something in a math book, it's been proven true.

Ancient cultures from all over each provided something. Several cultures came up with the same concepts independently, and others passed concepts on to other cultures. Historians don't know for sure exactly who did what when, because trustworthy history is based on written records, which don't always exist.

The following list gives you an overview of some of these historic contributions. There's no mystery in this very brief history — just highlights, folks, because the full story is enormous, and some of the details come up in the other chapters of this book. The point of this cultural timeline is to show that the math has been developing for a long time. The concepts in this book have been used by billions of people.

- **Prehistory** marks the birth of counting and tally sticks.

- **The Babylonians** introduced arithmetic, algebra, and geometry around 3,000 BC. Math not only was handy for measuring the farmers' fields but also helped the king collect taxes and astronomers look at the stars.

- **The Egyptians** gave humanity measurements, the math for agriculture (as early as 5,000 BC), and the math to build 138 pyramids (as early as 2,630 BC).

- ✔ **The Indus Valley civilization** produced the concept of the decimal system and the concept of zero (about 100 AD).

- ✔ **The Greeks** provided, among other things, the systematic study of mathematics (between 600 BC and 300 BC). That includes rigorous arguments and proofs.

- ✔ **The Romans**, among their many contributions, developed the standards still in use today for the weight and purity of gold and for precious metals and gemstones.

- ✔ **The Arabs** were the conduit for the discoveries of China and the Indus Valley civilization to Europe. They formalized the concept of zero and made other brilliant discoveries on their own.

- ✔ **The Chinese** developed math independently and were making strides as early as 300 BC.

- ✔ **The Europeans** produced some wonderful math before and during the Enlightenment. Isaac Newton gets credit for calculus. Copernicus gets credit for modern trigonometry, and René "I think, therefore I am" Descartes had many hits on the top-100 math charts. He's especially known for Cartesian coordinates. Head to Chapter 14 to see more about Cartesian coordinates.

- ✔ **The Mayans** had a super calendar and an excellent number system. They also had the concept of zero.

- ✔ **All the other civilizations** surely made unknown contributions. Where the historical record stops, the mystery begins. Perhaps the Hittites originated the credit card, or the Celts first developed the subprime mortgage.

The Trades, They Are A-Changing

Building technologies are probably the oldest trades, and they're in no danger of disappearing. In fact, they're more complex than ever. In addition, new trades are popping up regularly. As new careers come into existence and old careers evolve, the education and technologies that go with them must adapt. Trade schools (community colleges, technical colleges, and regional occupational programs) continue to offer vocational programs that expand as society's needs expand, including creating green programs as environmental consciousness becomes more socially important.

These expansions reflect three broad trends. First, once-new technologies, such as automobiles and air conditioning, are a permanent part of modern life (yes, Virginia, cars and AC weren't always common), so society must educate people to handle them. The second trend is the need to provide more specialized education for practitioners. The third trend is that education must offer training for the newest careers.

Here's a tough assignment (NOT): Go see a movie. Make it an animated one if you can. At the end of the film, study the credits as they crawl by — you see the names of dozens of traditional Hollywood specialized crafts, but you also see many new careers. Some skill areas, such as computer generated imagery (CGI), were absolutely unknown not so long ago.

Technical careers continue to evolve. Some jobs haven't been created yet, so we can't exactly list them here. Other jobs are turning into professions right now as the required skills become more formalized and people need more advanced education in how to do them. As technology advances, your career will likely evolve into something other than what it is now. If you maintain your current skills — especially your math skills — and keep your eyes open for what a new career requires, you can transition with no problem.

Math Devices That Can Help You Do Your Job

Specialized calculators and measurement tools help you do your work more efficiently, but that doesn't mean you should ignore general math skills. General math skills are great because they are *general*.

General skills are in your brain, which is a handy place for them. And you never have to replace batteries. You can use the skills in more than one career, which is excellent because experts say that the average person changes careers several times in a lifetime. Lastly, unlike a lot of tools you use on the job, you take your math skills home or anywhere else.

That said, some specialists make the same kind of calculations all the time, so specialty calculators devoted mainly to calculations needed for a particular trade are great additions to their math skills.

Some careers require more measurement than calculation; as a result, you can also find special measurement devices that give you exactly the information you need. Because some of them contain a computer chip (such as a nursing assistant's body mass index calculator), the machine does both measurement and calculation automatically, and the technician just sees the result as a measurement. The following sections give you a look at some of these calculators and measuring devices.

Pocket (or phone, or computer) calculators

Although your mind is an excellent calculator, life is short and some calculations are long. For complex math operations or any math operation on items with many digits you probably need a calculator.

Get off your wallet and buy a pocket calculator

A *basic* calculator (sometimes called a *four-function calculator*) is really simple and very inexpensive. This calculator does four basic math functions: addition, subtraction, multiplication, and division. But even the simple ones often include percentage and square root functions.

A more complex calculator is called a *scientific calculator*, but you don't have to be a scientist (or play one on TV) to use it. This calculator not only does basic math but also has more buttons so you can do trigonometry functions, exponents and roots beyond square roots, and logarithmic functions that use both base 10 and base e.

Both types are still called *pocket* calculators. They have come a long way, since the first ones ranged in size from a box covering your entire desk to a "handy" unit the size of a large book. Eventually, they shrunk in size to fit in your pocket, and some are now so small they fit on a keychain.

You know your cellphone has one

That's right! Your new mobile phone has a good calculator. For that matter, so does your older mobile phone. Just look for your phone's application icon or go to the standard menu to look for the built-in calculator. It's there.

Apple iPhones have a really cool calculator. When you hold the phone vertically (portrait mode), it's a "regular" four-function calculator. When you rotate it to the horizontal position (landscape mode), it becomes a scientific calculator.

Do it with a mouse

Desktop and notebook computers have come with built-in simple calculators for a long time. In the Microsoft Windows operating system, the classic version was a four-function calculator, as shown in Figure 2-1.

Figure 2-1:
Microsoft
Windows
four-
function
calculator.

Later versions of Windows have both a basic and scientific option. The scientific option is shown in Figure 2-2.

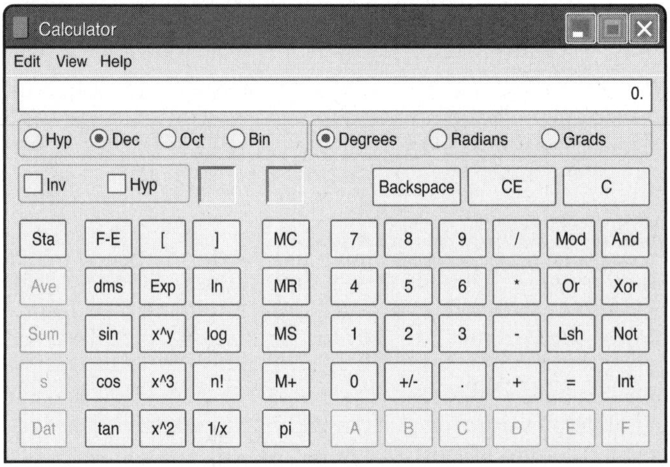

Figure 2-2:
Microsoft
Windows
scientific
calculator.

To run the calculator application in Windows, click Start→All Programs→ Accessories→Calculator.

The classic Apple Macintosh (Mac) calculator is a four-function calculator. The dashboard calculator *widget* (an on-screen mini-application) that comes with the Mac OS X operating system has three options: basic, scientific, or programmable.

Spreadsheet programs are your friends

Microsoft Excel is the spreadsheet software that's been wildly popular on PCs and Macs. It does far more than just calculations, but the calculations alone are impressive — the program contains dozens of built-in functions.

If the cost of Excel (about $230) concerns you, look at Sun Microsystems' OpenOffice.org (billed as "the free and open productivity suite"). The Calc spreadsheet program is very similar to Excel, and the price is $0. Find it at www.openoffice.org.

Specialty calculators

Specialty calculators use predetermined formulas, and that's okay. After you know the math behind the formulas, you can use the formulas with confidence. In the following sections, check out some of the many specialty calculators you can find. To get you started, you can find an extensive set of online calculators (hundreds of them) at www.martindalecenter.com/Calculators.html.

Sometimes an Internet calculator isn't really a calculator — it's a table of factors to consider in doing a calculation. Such calculators work, but you're better off to consider them estimators rather than calculators.

Machinist calculator

Machinist calculators are geared (no pun intended) for machine shop calculations. You can find numerous machinist calculators online, including the Trades Math Calculator for PCs. Go to www.tradesmathcalculator.com/ or www.freedownloadmanager.org/downloads/machinist_calculator_software/.

One of the great machining calculators isn't a calculator at all. It's *The Machinery's Handbook,* a giant book of tables (the 2008 edition — the 28th — is over 2,700 pages!). It was first published in 1914.

Conversions and one-time calculations

The Internet is filled with free, simple conversion programs. You can find conversions for angles, weight, temperature, fuel consumption, and so on. For example, point your browser to www.onlineconversion.com/.

Plumbing and pipefitting calculators

Flow calculations are important in many plumbing applications, and a pipe flow calculator is a great example of the computerization of complex formulas. This calculator helps you work out pipe pressure drop and pipe diameters. Visit www.pipeflowcalculations.com/.

Roof surface area calculator

For roof coatings, one estimating calculator uses a table. You start with the interior square footage of the house to be roofed. Then you make allowances for roof overhang, the thickness of the walls, and the slope of the roof. After that, you consider the type of roof to be coated, as well as wastage. When those figures are all in place, the table gives you the total roof surface area to buy coatings for. You can find this tool at www.somay.com/roof_coatings/ roof_calculate/roof_calculate.html.

You can also work out roof surface area from direct measurements or from blueprints, and that may be the smarter way to go because an online roof surface calculator can only provide an approximation.

Thermometers and sphygmomanometers

A key part of the certified nursing assistant's (CNA's) work is to take a patient's vital signs — temperature, blood pressure, and pulse. Thermometers are essential for taking the temperature of the human body. For decades, the temperature was displayed as the height of a thin column of mercury in a glass tube. Now, digital thermometers display temperatures in either degrees Fahrenheit or degrees Celsius.

Blood pressure is measured with a blood pressure cuff, or *sphygmomanometer*. That's easy for you to say! Try saying "*sfig*-mo" and combine it with "man-*om*-eter." That's the device with an inflatable cuff and gauges. Figure 2-3 shows you an old-fashioned version, but these days, many blood pressure cuffs are digitized, and the CNA can take the readings directly from the instrument. The blood pressure cuff measures systolic and diastolic blood pressure.

Blood pressure measures two kinds of pressures in the arteries, which is why it includes two numbers. A measurement such as 120/80 comes from old blood pressure devices that showed the pressures as the height of columns of mercury in two tubes. So, *120* refers to a column of mercury pushed up to 120 millimeters by the patient's systolic blood pressure, while the *80* refers to the patient's diastolic blood pressure. For more on working with metric units, see Chapter 6.

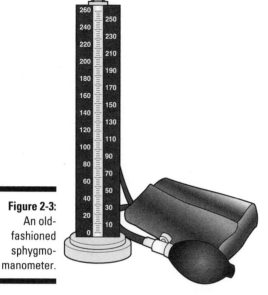

Figure 2-3:
An old-fashioned sphygmo-manometer.

The CNA determines your pulse rate by counting just a few heartbeats while watching a watch for a few seconds; she then multiplies to get beats per minute. Pulse is traditionally measured by using a stethoscope and a watch. However, today's digital blood pressure cuffs include pulse rate in the displayed output.

Micrometers, calipers, and gauges

Machinists have incredibly complex jobs, making the parts for just about every item you use. Even parts made of plastic are machined, or the molds to make them are. At its simplest, a machinist does drilling, milling, turning, and grinding. Except that's not simple.

Nowadays, most machine tools (such as mills, turning centers, and drilling machines) are computerized. The machine does much of the cutting for the machinist, but that doesn't make things automatic. Every step has to be programmed, and the machinist has to calculate feed rates and cutting speeds, based on the material and the cutting tools used.

Online and downloadable calculators can help machinists with the needed calculations. (Check out the earlier section "Machinist calculators" for information on these calculators.)

When the work is complete, the machinist uses micrometers, calipers, and gauges (shown in Figure 2-4) to make sure the work is accurate. These are measurement tools, not math tools.

Micrometers are also common in engineering applications wherever you must measure small dimensions precisely. You see calipers used in a variety of fields, including automotive technology (see the example in Chapter 15) and medicine.

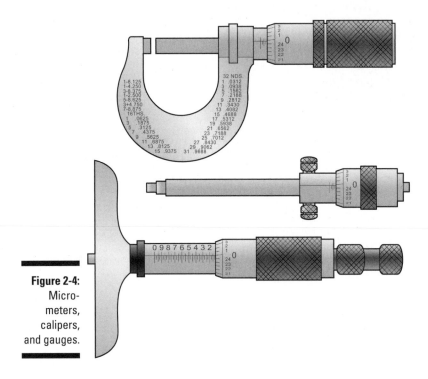

Figure 2-4: Micrometers, calipers, and gauges.

Automotive tools

The local garage says that its mechanics can acquire up to $50,000 in tools after just a few years working. Many are specialty tools, such as the gas analyzer in Figure 2-5, used only in auto work. Devices like brake thickness gauges, coolant testers, and hand code readers require measurement and recording information, not math.

One thing auto technicians won't want to acquire on their own is a *diagnostic computer,* which analyzes the various computer systems in a vehicle. Because so many cars are managed by one or more computers, it takes a shop diagnostic computer (with frequent upgrades) to be able to read the codes that cars generate.

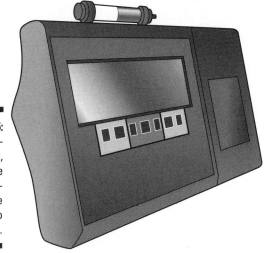

Figure 2-5:
Many auto-
motive tools,
such as the
gas ana-
lyzer, are
unique to
mechanics.

The *smog check computer* (formally, an *emission analyzer*) is another computer at the auto mechanic's shop. It determines whether your car meets
your state's emission standards. After a number of visual checks (for example,
fuel cap, crankcase smoke), the mechanic connects an RPM measurement
device and puts a probe in the tailpipe. The computer does the rest, including
sending the results over phone lines to the Department of Motor Vehicles.

In a world with that much automation, it's kind of a comfort to know that you
still measure cylinder bores (diameters) and brake pads (thickness) with
manual tools (calipers and brake thickness gauges, respectively; see the preceding section for more on calipers).

Tire pressure is a common measurement at the garage, with digital gauges
often replacing the older mechanical gauges.

One great retro tire measurement that doesn't even involve math is the penny
test for tire tread depth. Place a penny into several tread grooves on the tire,
with Lincoln's head pointing in. If part of his head is covered by the tread, you
have more than $\frac{2}{23}$ inch of tread depth remaining. ***Note:*** Legal tread depth is
defined in 42 states as $\frac{2}{23}$ inch, with some variations in other states.

Carpentry tools

As a carpenter, your work is measurement- and calculation-intensive. That's
no surprise. What is a surprise is that, with all of the computerized measuring
tools available, the carpenter's tape measure is probably the most frequently-
used tool of the trade. You use it to make measurements and also to apply

the results of your calculations to wood that's to be cut or drilled. The math involved is usually arithmetic (addition, subtraction, multiplication, and division), which we cover in Chapters 4 and 5.

The *steel square* may be a contender for second place. That tool helps you lay out right angles, but it's more important for laying out rafters, hip rafters, and stairs. When you lay out a roof pitch, the math is arithmetic, with a bit of ratio work thrown in. (See Chapter 8 for more on ratios.) A great low-tech carpenter's tool is the *spirit level* (which you probably know as just the *level*). It's a short length of wood or metal with a small liquid-filled tube in the middle. The tube contains a bubble; when the level is horizontal, the bubble rests between two marks. Figure 2-6 shows you an example. High-class spirit levels may have additional tubes for measuring vertical or 45-degree inclinations.

Figure 2-6:
A level.

The level has pretty much resisted going high tech. However, newer laser levels project straight lines with the touch of a button. What makes the line level with this device? Yes, a little built-in bubble spirit level.

Bricklaying tools

Bricklayers need their work to be level and straight. It's essential and (along with mortar throwing) is part of why bricklaying is an art form. Chalk lines (made with spirit levels such as the one shown in Figure 2-6 and a tool called a *chalk line,* shown in Figure 2-7) are important. Math is secondary in the basic work; the important thing is to use the tools correctly.

However, as a bricklayer, you need to use a relatively large amount of math in preparing for work. You use multiplication to determine the area of a wall and division to calculate how many bricks the job requires. This figure in turn leads you to calculate how many bags of cement and sand (or pre-mixed mortar) you need for the job. See Chapter 5 for details about multiplication and division.

Figure 2-7:
A chalk line.

Chapter 3

Zero to One and Beyond

*N*umbers aren't only part of civilization but were in use before civilization even existed. With the earliest peoples, you can imagine that verbal communication wasn't even necessary when it came to mathematics. If Og found some mastodons for dinner, he went to his tribe, made the sign for a mastodon, and pointed in the direction where he saw them. Then, even though he had hoped the group would hold their questions until the end of the presentation, someone would jump in and make the sign for "How many Hairy Tusk Beast you find?" Og had the answer "at hand." He would hold up some fingers to indicate how many, and the gang would run off to hunt mastodons.

Og's descendants now live in an era of double-knit stretch polyester and smartphones, but the need to use numbers for communication hasn't changed. Some careers are more number-intensive than others, but every trade uses numbers. If you have friends who say they can't do math, please remind them that they *can* do numbers. This ability is what separates human beings from the lower-order creatures, such as oysters and fire hydrants.

The beauty of numbers in counting (their simplest application) is that answers come with no skills besides counting. However, even counting requires careful administration. And for speed and efficiency, you can go beyond counting to arithmetic, as we show you in later chapters.

In this chapter, you review the common types of numbers you work with and some uncommon, strange, and unbelievable numbers, too. You also explore the secrets of zero. All this requires no more than a set of fingers and toes to count with.

Looking at the Numbers that Count: Natural Numbers

Natural numbers are basic numbers, which are also called *counting numbers.* Most people just call them *numbers.* Natural numbers have a familiar look: 1, 2, 3, 4, 5, 6, and so forth. They're *whole* numbers (as opposed to fractions) and they don't include zero (0) or negative numbers. They serve two purposes:

- ✓ **Counting:** *Counting* is the technique you use for inventory and all stock keeping. Natural numbers are also the fundamental unit of purchasing, no matter what your line of business is. Today, online shopping carts ask the fundamental counting question "How many?" Whether you're counting or buying 500 milliliter Erlenmeyer flasks or barrels of transmission fluid, you've got to know what quantity you have or want. Using natural numbers takes on a personal meaning after work. As you stand in the express checkout line, your blood is chilled by a sign reading "Ten items or less." You must quickly count the items in your cart (notice that the types of products, their prices, or their sizes don't matter anymore) to make sure you're not above the maximum.

- ✓ **Ordering:** You use natural numbers for *ordering,* describing things in a certain order. When you list the first (1st), second (2nd), third (3rd), and fourth (4th) largest cities in your state, you're using natural numbers for ordering. In your personal life, using numbers for ordering becomes painfully clear at the Department of Motor Vehicles. You hold a small piece of paper that says "#89," and an electric sign says "Now serving #4." The numbers show the order in which people are being served and your position in that order. They also show that you're in for a long wait.

The set of natural numbers doesn't include zero. In simple counting, you can't have zero apples or zero oranges. Zero is part of a larger group of numbers. Check out "Zero: Making math easier" later in the chapter for more on this number.

However, an exception exists in the field of computing: Zero becomes the first counting number and takes the first position in arrays and other data structures. Don't be surprised to see *for(i=0;i < 100;++i); sum = sum + grades [i]*; used to loop through positions 0 to 99 in a 100-element array.

Integers: Counting numbers with extras

Integers are like counting numbers, but there are more of them. The set not only includes the counting numbers (1, 2, 3, 4, 5, 6, and so forth) but includes zero (0) and negative numbers (–1, –2, –3, –4, –5, –6, and so forth). You can also call these numbers *whole numbers*. Say the word *integer* with a soft *g.* That is, say "*in*-tuh-jer." Taken together, integers form a nice line:

$$\ldots -6\ -5\ -4\ -3\ -2\ -1\ 0\ 1\ 2\ 3\ 4\ 5\ 6 \ldots$$

Integers can be positive or negative, odd or even. A *negative number* is a number that's less than zero. Of course, a *positive number* is greater than zero. An *even number* can be divided by 2 with no remainder. An *odd number* can't be evenly divided by 2. Zero is an even number. However, it's not positive or negative — it's just zero.

These integers look as orderly and evenly spaced as the chorus line in a Broadway musical. And it's no wonder, since each integer differs from the others beside it by just one. As you can imagine, if you have an infinitely wide stage, the negative integers at the left extend forever. The positive integers at the right do the same thing.

Integers are the numbers you use to perform all simple math. You can do all arithmetic operations (addition, subtraction, multiplication, and most division) with integers. Integers are also useful for plotting the points on a graph or chart.

Where do they get these names? The word *integer* comes from Latin and means *untouched.* You can't touch an integer, so you can't break it — it's an unbroken or whole number. (Speaking of untouchable, *integer* is a relative of the word *integrity.*)

Zero: Making math easier

What is zero (0)? How can it be important when it's really nothing at all? Zero may look like nothing, but it represents something — it appears in numbers and in calculations where digits ought to be. It's a *placeholder,* a kind of punctuation mark that helps you interpret numbers correctly.

And why's that valuable? Because you (and most of the world) use a decimal number system, and it's a *positional* system. In a three-digit number, such as 123, those digits are more than just a *1,* a *2,* and a *3:*

- The position of the *1* means that 1 is the number of *hundreds,* because it's in the third column from the right.

- The position of the *2* means that 2 is the number of *tens,* because it's in the second column from the right.

- The position of the *3* means that 3 is the number of *ones,* because it's in the first column from the right.

Pretty straightforward so far, because you don't need a placeholder. But what happens when you have one hundred and three single items? How do you write that without a placeholder?

- ✔ You can write 13, but that's misleading and just plain wrong.

- ✔ You can try 100 + 3, but such a system of notation makes math operations much tougher.

- ✔ You can try Roman numerals and write 103 as CIII, but the disadvantages of Roman numerals (with only seven symbols and no zero) are many and make the system a poor candidate for math.

Ladies and gentlemen, boys and girls, what you need is a placeholder. In the number one hundred and three, let the 1 show one hundred, use a 0 show no tens, and have the 3 show three units, giving you 103.

Zero and the decimal system made most other math systems obsolete. Mathematicians point out that the decimal system is a base 10 system. The Maya of Central America used a base 20 system, and they used zero, too. There are vestiges of the base 12 system in today's twelve-hour clocks. And for the nerdy band of brothers, the computer age brought forth the base 2 (binary) and base 16 (hexadecimal) systems.

Zero can be your biggest friend in mathematics because it makes for quick work:

- ✔ Whenever you multiply anything by zero, the answer is always zero. For example, $3 \times 0 = 0$; $274,561 \times 0 = 0$; and so forth.

- ✔ When you add 0 to a number, the answer is the same number. For example, $2 + 0 = 2$, $27 + 0 = 27$, and so forth.

- ✔ Any number raised to the power of 0 is 1. For example $756^0 = 1$, and $7^0 = 1$. See Chapter 11 for more on powers and exponents.

See how nice zero can make your math life? Anytime you're solving a math problem, look for zero. It doesn't look like much, but it can help you.

Where did zero really come from?

Historians cite many different civilizations that might have developed decimals and zero, but the system may have evolved in the Indus Valley (near the western edge of modern-day India), and the Indians may have gotten techniques from China. Then the word spread to the Middle East. In 976, Muhammad ibn Musa al-Khwarizmi said if there was no number in a place, you should use a little circle. The Arabic word for that little circle is *safira* (meaning *it was empty*) or *sifr* (meaning *nothing*). That led to the modern English word *cipher*. Sifr also leads us to the Italian word *zefiro* (meaning *zephyr* or *zephyrum*) and the Venetian contraction *zero*.

Going Backward: Negative Numbers

As we mention earlier in the chapter, a *negative* number is a number that's less than zero. You represent negative numbers with a minus sign; for example, –1, –23, and –8,542. Zero isn't negative (but it's not positive, either).

Negative numbers may seem like a fantasy concept, but they're very real in many lines of work.

Working with negative numbers

In mathematics, negative numbers are a concept. But concepts don't put bread on the table (unless you're a mathematician). In your daily activities, you work with negative numbers in the real sense, and they almost always represent a reduction or a deficit.

In some trades, when a positive quantity decreases, the math "stops" at zero. In parts management, food management, or hospital stockroom management, when you have 0 of something, you're all out. There's no concept of negative widgits, negative eggs, or negative IV solutions. But the reason you get to 0 units is because of inventory draws (reductions), and each reduction in inventory is the application of a negative number. Stock on hand minus the amount of the draw results in a new, lower amount of stock on hand.

In virtually all trades, accounting transactions can result in amounts lower than zero. For example, when a cosmetologist is sick, she has no clients (no inflow of cash), but the rent is due on the station at the salon (outflow of cash). Not so good. Low income and high expenses can occur in the construction trades, the automotive trades, and even in a doctor's office. Negative cash flow is a real and painful concept. And if the business checking account is overdrawn, that's a very serious negative number.

Negative numbers aren't always grim. For example, the countdown — the process of counting hours, minutes, and seconds backward until something happens — is a "positive" application of negative numbers. It's part of NASA rocket launches, adds drama to action movies, and announces the start of each new year.

Traveling down the number line

In mathematics, negative numbers are part of series of numbers. One way to visualize the series is to draw a number line. Put 0 in the center, mark positive numbers to the right of 0, and mark negative numbers to the left of zero. Figure 3-1 shows a number line.

Figure 3-1:
A number
line.

The farther to the right of 0 on the number line you go, the larger the numbers get in value. The farther to the left of 0 you go, the more the numbers decrease in value. Looks can be deceiving. For example, the *9* in –9 is has a larger *magnitude* than the *8* in –8, but the minus sign (the negative sign, –) makes a difference. A larger negative number has less value than a smaller negative number.

And although negative numbers may seem to be the opposite of positive numbers, they act the same, and you can do the same math operations with them.

Getting Between the Integers: Fractions, Decimals, and More

Life was simpler in the third grade with only integers to deal with. But then again, you didn't get a paycheck for attending the third grade. So there comes a time in your career when you must also know about other types of numbers.

In between the integers are many other numbers, known as common fractions, decimal fractions, rational numbers, and irrational numbers. Although integers are nice, clean numbers to work with, you can't ignore the numbers in between them.

Our fractional friends

Fractions are the most common numbers in the technical careers. A *fraction* is part of a number, more than zero but less than one. The word comes from the Latin *fractus* or *frangere,* which means *broken* or *to break,* as opposed to integers, which are unbroken whole numbers.

The two kinds of fractions are *common fractions*, which look like this:

$$\frac{1}{2}, \frac{7}{9}, \frac{27}{651}$$

and *decimal fractions*, which look like 0.46, 0.375, or 0.87695. If you combine a whole number with a fractional number, the result is called a *mixed number*. For example, 5.243, $14.95, or

$$3\frac{4}{5}$$

Check out Chapters 8 and 9 for more on fractions.

Figure 3-2 shows how fractions on the number line fall between the integers.

Figure 3-2:
Fractions on
the number
line.

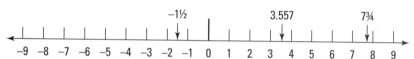

The leap from 0 to 1 is "one small step for math, one giant step for mathkind." It looks small, but in between 0 and 1 are many fractions (an infinite number, as it turns out).

The rational numbers (and their irrational friends)

On the job and in your personal life, you have two kinds of friends: rational and irrational. Both kinds are valuable to know (except maybe the one who puts bean sprouts and peanut butter on pizza). The same is true with rational and irrational numbers: You're better off knowing both.

A *rational* number can be expressed as a *ratio,* the quotient of two integers. Any common fraction fills the bill, as it shows the ratio of the top number to the bottom number. What about 0.75? This decimal number is really the fraction seventy-five one hundredths, and when shown as a common fraction:

$$\frac{75}{100}$$

you see that it's a ratio.

Following are examples of rational numbers:

$$1 \quad \frac{3}{2} \quad -.5 \quad \sqrt{4}=2 \quad \sqrt[3]{27}=3 \quad 3.54982$$

Like the integers and like the fractions, mathematicians have proved that an infinite number of rational numbers exist.

You can express some numbers as fractions, but they produce infinite decimals in a repeating sequence. For example:

$$\frac{1}{3} = 0.33333\ldots \text{ or } \frac{1}{11} = 0.09090909\ldots$$

An *irrational* number is always acting out. It won't let you express it as a simple fraction, and as a decimal fraction the digits go on forever in no repeating sequence. As you see, the irrational numbers don't behave in a rational way.

The most famous irrational number is π (pi, the Greek letter, pronounced *pie*). Pi the ratio of the diameter of a circle to its circumference. You were first exposed to π in grade school, and you may use it in work if you calculate circular areas or the volumes of cylinders.

Pi has been calculated to over one trillion (!) decimal places, and the calculations still don't come out evenly. And they never will.

At least once, the government has tried to legislate the value of pi. (Oh, how we authors wish we made these things up.) In 1897, the Indiana House of Representatives considered a bill that would have set pi to a value of 3.2.

Two other numbers, Euler's Number (the number *e*) and the Golden Ratio (represented by the Greek letter phi, ϕ), are also famous irrational numbers.

Taking a Look at the Lesser-Known Numbers

The numbers you encountered so far in this chapter are the numbers you use in your work and at home. Here is the lightning round of other number types. One type describes everything we've discussed in the chapter so far, two types are never used except by mathematicians, and (to finish on a positive note) you use the last two types every day.

Real numbers

Real numbers is the name for all the numbers covered in this chapter to this point. That includes natural numbers, integers, fractions, positive numbers, negative numbers, zero, rational numbers, and irrational numbers. They're all real, meaning you can find any of these numbers somewhere on the number line.

Imaginary numbers

An *imaginary number* is a number that includes the square root of −1. This value is supposedly impossible. In real life, you can't square a number and get −1, but in conceptual life, you can. The math expression is:

$$i = \sqrt{-1}$$

The symbol for the imaginary unit is *i*. A number that includes *i* (for example, *i* or 7*i* or −3*i*) is imaginary.

Early mathematicians thought imaginary numbers were useless. In the 1600s, mathematician René Descartes wrote about them. He used the term "imaginary," and he didn't mean it as a compliment.

But the world of mathematics evolved, and in time mathematicians found the concept of imaginary numbers to be very useful. You use imaginary numbers in engineering disciplines like signal processing and vibration analysis.

Complex numbers

These aren't numbers with a psychological disorder. But they're not simple numbers, either. A *complex number* is a combination of a real number and an imaginary number.

You write it in the form $a + bi$, where *a* and *b* are real numbers, and *i* is the standard imaginary unit.

Nominal numbers

A *nominal number* (sometimes called a *categorical number*) is a number you use for identification only. It doesn't matter what the value of the number is. Here are some examples of nominal numbers:

- Social Security Number
- Vehicle Identification Number
- Drivers License Number
- Inventory part numbers
- Universal Product Code
- The combination for a lock or safe. (Some locks and safes even use letters instead of numbers.)

Indefinite and fictitious numbers

You'd think *number* means *precise* and *precise* means *number*. No, not so. Humans are an infinitely clever species. They created the pet rock, the Furby, and the iPod. And indefinite numbers.

An *indefinite* number is a term for a quantity. You use it when you don't know the exact amount of the quantity. The following are just a few indefinite numbers. You know, just a handful.

- A few
- Several
- Plenty
- Scads

- Bunch
- Many
- Lots
- Heaps
- Tons
- Buckets

These are "quantity" indefinite numbers. "Amount" indefinite numbers include *mite*, *smidge*, *glug*, and *dollop*.

Handling Numerical Story Problems

Story problems (sometimes called *word problems* or *life problems*) are math problems where the details are presented more as a story than with straight figures. Not to worry. Story problems aren't hard to solve. Check Chapter 7 for all the details about story problems.

Example: Automotive tech — a slippery task

You work at a BMW motorcycle dealership. You hope to study more automotive technology in school and open your own shop some day. But at the dealership, you have a pretty basic starter job, and your boss asks you to determine the on-hand quantity of BMW motor oil. Determine how many plastic containers of oil you have.

1. **Take a look at the entire quantity to be counted.**

   The following figure shows the number of containers to be counted:

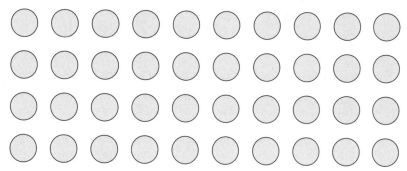

2. **Count the containers.**

   The answer is 48 containers.

   The boss is pleased. He takes you to another stockroom and gives you a similar problem. The following figure shows a new group of containers to be counted:

Simple counting solves the problem. In the figure above, you have 448 containers.

Arithmetic, however is more efficient. With arithmetic, you would have figured out there are 16 rows of containers, with 28 cans in each row. Then you would have used multiplication to multiply the number of rows by 28 to get the total number of cans.

The moral of the story (and actually there are two):

✔ Counting is good, but it has its limitations.

✔ Doing a good job on a tedious task sometimes gets you involved with a *more* tedious task.

By the way, BMW motorcycles don't use oil — they use *engine lubricant*. The difference between that and oil? About $4 per quart.

Example: Getting the order right

You and a co-worker are in the drive-through at Taco Palace. Together, your order is three tacos, two burritos, and two soft drinks. When you get to the pick-up window, you get your order but it doesn't seem right. Use counting to confirm its correctness.

1. **Count the soft drinks.**

   Either speak the number or use your fingers. If you speak, the results should be "One, two."

2. **Count the burritos.**

   Again, either speak the number or use your fingers. The results should be "One" "Two."

3. **Count the tacos.**

   "One" "Two"

Wait! Something's wrong here. You ordered three tacos and you only got two. The order is short. Let the person at the pick-up window know.

Notice that for this problem only counting was required. The problem doesn't need arithmetic. The beauty of numbers in counting (their simplest application) is that answers come with no skills beside how to count. Although this example is extremely simple, it shows that counting is sometimes the fastest, most accurate way to solve a problem.

Chapter 4

Easy Come, Easy Go:
Addition and Subtraction

*E*very career has its basics — fundamental techniques that are the starting points for every other technique. In carpentry, you begin by ham-
mering a nail and making a simple cut with a saw. In cooking, you begin with
simple recipes. Even in sports, such as boxing, baseball, and the martial arts,
you start with a basic stance. In fact, no matter what you're doing, if you
don't get the fundamentals right, you're pretty much assured of having problems later.

Math is the same way. The basics are addition and subtraction. They're just
a step beyond counting and form the basis for all other math operations.
Counting is well and good (see Chapter 3), but eventually you run out of fingers and toes. That is to say, eventually counting is tedious, and you must go
on to addition and subtraction.

If you live in the rain forest (like, for example, the Nadëb of Nadahup, Brazil)
and don't have any commerce, you may have no need to count, add, or subtract. Your number system doesn't have any specific numbers above one. The
Nadëb use a "1-few-many" system.

Because you probably don't live in the rain forest, you should read this chapter. It's not one page long. It's not many pages long. It's a few pages long.

In this chapter, you review exactly what addition is and the parts of the addition operation. You do the same thing with subtraction. These two operations don't have much mystery, but you explore what mysteries they do have — carrying in addition and borrowing in subtraction. You also see the easiest ways to check your work.

Making Everything Add Up

Addition is the process of combining quantities. You probably knew this, because addition is an operation you grow up with. Mothers command their sons, "Johnny, tell Aunt Ida how much one and one is." And sons (even if their names aren't Johnny) respond by saying, "One and one equals two." This exchange is cute when you're 3 years old, but by the time you're 21, it's tiring for both you and Aunt Ida.

Each item to be added is an *addend*. The result is the *sum* (which comes from the Latin *summare*, "the highest").

When you add, use a plus sign (+). You add in a row and use an equal sign (for example, $1 + 2 + 3 + 4 + 5 = 15$) or in a column using an underline. For example:

$$
\begin{array}{r}
1 \\
2 \\
3 \\
4 \\
\underline{5} \\
15
\end{array}
$$

Addition works with all kinds of numbers — integers, zero, rational numbers, fractions, and irrational numbers. In fact when you see a mixed-number such as

$$3\frac{1}{2}$$

that's really the addition of 3 and $\frac{1}{2}$.

That is:

$$3+\frac{1}{2}=3\frac{1}{2}$$

The same is true with decimals. For example, 3.654 is the sum of 3 and 0.654. That is:

$$3 + 0.654 = 3.654$$

Adding numbers in a column

When you have several numbers to add (say, more than three), the best way to add is in a column. To add in a column, simply arrange the numbers so that they are all aligned on the right side, and begin the addition process.

$$
\begin{array}{r}
553,141 \\
221 \\
3 \\
45,454 \\
17 \\
\hline
598,836
\end{array}
$$

This technique is obvious in a spreadsheet such as like Microsoft Excel or OpenOffice Calc. Because these applications are cell-based, vertical columns are the logical approach to addition. Figure 4-1 shows addition in Excel.

	A	B
1	553,141	
2	221	
3	3	
4	45,454	
5	17	
6	598,836	
7		

Figure 4-1: Adding numbers in Microsoft Excel.

Note that you can also add cells horizontally in a spreadsheet. This feature is handy when you want to verify addition of a large number of items in several rows and columns. The technique is called *downfooting and crossfooting*, and you can find details on the Internet. Basically, you sum the columns and the rows and compare the results for accuracy.

Be careful with certain tools. A spreadsheet shows you what items you're adding. So does a printing calculator (yes, they still make them). But when you use a pocket calculator or smartphone, you usually can see only the growing sum "so far" in the LCD window. Be sure that you've entered all the items to be added if you want the correct sum.

Adding zero

When you add 0 to a quantity, the quantity doesn't change. For example, 23 + 0 = 23. You can do this all day long (for example, 23 + 0 + 0 + 0 + 0 + 0 = 23), but that won't change the answer.

And of course, adding 0 to 0 gets you 0. You express this as 0 + 0 = 0. This concept is what's called, in technical terms, a whole lotta nothin'.

Adding negative numbers

If you had the wisdom, taste, and discernment to read Chapter 3, you know that negative numbers have a minus sign (–) in front of them and are numbers less than 0. (Don't worry; you can still get wisdom, taste, and discernment by reading Chapter 3 later.)

You can add negative numbers. No problem. The result of adding negative numbers is a larger negative number.

In algebra, you express the idea as:

$$-a + (-b) = -c$$

Adding multiple negative numbers is as easy as adding multiple positive numbers. For example, –1 + (–2) + (–3) + (–4) + (–5) = –15. It's clearer to illustrate this addition as a column, because it eliminates all those + signs.

$$
\begin{array}{r}
-1 \\
-2 \\
-3 \\
-4 \\
\underline{-5} \\
-15
\end{array}
$$

Carrying the extra

Adding in any digit position (for example, the ones column, the tens column or the hundreds column) is easy. But what do you do when the result it more than ten? You carry.

The term *carry* means that when the results of adding a column are higher than 9, you record the right-hand digit (the ones number) and add the left-hand digit (the tens number) to the next column.

For example, if you add 5 + 4 = 9, it is (in professional terms) called a no brainer. Just add.

But what about adding 428 and 186? Not so simple.

```
 428
 186
 ???
```

Adding the 6 and the 8 in the ones column gives you 14, but the ones column can't hold more than a single digit. (Well, to be fair to the poor ones column, no column can hold more than a single digit.) What to do?

Record the 4 and carry the 1 to the next column. This process is simple, but we describe it fully here anyway for clarity. Look at the addition of 6 and 8 in the ones column.

```
  1
 428
 186
 ??4
```

The sum of 6 and 8 is 14, so you write the 4 as part of the sum and carry the 1 to the top of the tens column. Then it's "second verse, same as the first." Add the carried 1 to the 2 and the 8 in the tens column.

```
 11
 428
 186
 ?14
```

The result is 11, and that two-digit result won't fit in a column that was built to hold one digit, much like those pants you were able to fit into before Thanksgiving and Christmas dinners but can't now. So, you write the 11 as 1 in the tens column and carry the one to the top of the hundreds column.

Now, one more addition completes the work.

$$\begin{array}{r} 11 \\ 428 \\ 186 \\ \hline 614 \end{array}$$

The answer is $428 + 186 = 614$.

Checking your work

This tried-and-true tip is old as the hills and twice as dusty. To check addition, add the column in reverse order. You should get the same sum. If the problem is

$$\begin{array}{r} 553,141 \\ 221 \\ 3 \\ 45,454 \\ 17 \\ \hline 598,836 \end{array}$$

Just add "up" the column.

$$\begin{array}{r} 17 \\ 45,4554 \\ 3 \\ 221 \\ 553,141 \\ \hline 598,836 \end{array}$$

The answer should be the same. You can also subtract the individual addends, one at a time, from the sum until there's 0 left, but that's a far more tedious process.

Subtraction: Just Another Kind of Addition

Subtraction is the process of removing a quantity from a quantity. It's also called the *inverse* (or the opposite) of addition.

Subtraction is the second math operation you grew up with (besides addition). Your Aunt Ida may have said, "Jane, if you had three apples and I took two, how many would you have left?" The mathematically and politically correct answer is, "One apple, Aunt Ida." The other answer — the one that will get you in trouble — is, "I'd still have three, you greedy old woman, because I wouldn't let you take any. And you'd have a black eye for trying to steal my apples."

Subtraction is the only tool for figuring differences and remainders. The words essentially mean the same thing and you may see either of them in subtraction problems. A *remainder* is the balance of a quantity left after it has been reduced by subtraction. A *difference* is a numeric comparison between two quantities.

- ✔ If you have five apples and give two apples away, how many are left? The number you have left is the remainder.

- ✔ What's the difference between $600 and $800? The answer, $200, is a comparison of two quantities.

- ✔ If you see a mileage sign, "Omaha, 180 miles" and the last one you saw said "Omaha, 210 miles," how far have you traveled in between them? The number of miles you've travel is the difference between the first and second distances.

The number you subtract from (which is usually the larger number) is called the *minuend*. The number you are subtracting is the *subtrahend*. The result is the *difference*. Rest assured, these terms don't come up very often in everyday talk.

When you subtract, use a minus sign (–). You can subtract inline and use an equal sign (for example, 23 – 10 = 13) or you can subtract in a column. A subtraction problem in a column looks like this:

$$\begin{array}{r} 23 \\ -10 \\ \hline 13 \end{array}$$

Like addition, subtraction works with all kinds of numbers — integers, zero, rational numbers, fractions, and irrational numbers.

Subtracting a positive is the same as adding a negative

What does it mean to say that subtraction is the inverse of addition? In algebra, you'd say

$a - b$ is also $a + (-b)$

Yes, subtracting b from a is the same as taking a and adding $-b$ to it. This may not seem like a big deal now, but it becomes important when you get to algebra (see Chapter 12). For more info about positive and negative numbers, flip to Chapter 3.

Subtracting negative numbers

You can subtract negative numbers. The peculiar thing is that *subtracting a negative* is like *adding a positive*.

In algebra, you express the idea as

$-a - (-b)$

The catch is that subtracting a negative number changes the sign. The result is

$-a - (-b) = -a + b$

Subtracting zero

When you subtract 0 from a quantity, the quantity doesn't change. For example, $23 - 0 = 23$. Subtracting 0 from 0 is equal to 0. That's $0 - 0 = 0$, and that's nothing.

Subtracting multiple items

You can subtract multiple items all at once, but be careful — doing so may be a little confusing.

$$
\begin{array}{r}
23 \\
-10 \\
-5 \\
\underline{-5} \\
3
\end{array}
$$

This column of subtractions really represents multiple individual subtractions. $23 - 10 = 13$, $13 - 5 = 8$, and $8 - 5 = 3$.

Subtracting multiple items is more obvious in a spreadsheet. Figure 4-2 shows subtraction in Microsoft Excel.

Figure 4-2: Subtracting numbers in Microsoft Excel.

	A
1	23
2	-10
3	-5
4	-5
5	3
6	

The answer is 3, and it's really the result of adding all the numbers (positive and negative) together.

When you're subtracting multiple items, be careful. Spreadsheets and printing calculators tell you the whole story, but pocket calculators and smartphones only show the results so far. Be sure you've entered all the items to be subtracted if you want the correct answer.

You can take shortcuts, such as adding all the negatives first. In the example

$$
\begin{array}{r}
23 \\
-10 \\
-5 \\
\underline{-5} \\
3
\end{array}
$$

you can first combine the negatives.

$$
\begin{array}{r}
-10 \\
-5 \\
\underline{-5} \\
-20
\end{array}
$$

The subtractions are lumped-together negatives, and you subtract them from the positive amount:

$$
\begin{array}{r}
23 \\
-20 \\
\hline
3
\end{array}
$$

The answer is still 3, just as before.

Borrowing when you have to

In real life, you're better off if you avoid borrowing. In subtraction, you do it all the time to make subtraction easier.

The term *borrowing* refers to converting one unit from the next position (at the left) into the units you are working with. As you know, the positions of the numbers are the ones column, the tens column, the hundreds column and so forth.

You can freely borrow from the column at the left of the column you are working in. To illustrate this, say you want to do this subtraction:

$$
\begin{array}{r}
423 \\
-156 \\
\hline
???
\end{array}
$$

Look at the ones column. If you were subtracting 3 from 6, you could easily do it in your head. But how can you take 6 from 3? You can't, but you can take 6 from 13.

But where do you come up with 13? Look at the subtraction problem this way:

$$
\begin{array}{rrr}
400 & 20 & 3 \\
-100 & -50 & -6 \\
\hline
??? & ??? & ???
\end{array}
$$

To make the 3 in the ones column into 13, just borrow a single 10 from the tens column. That reduces the tens column by 10, and the problem looks like this:

$$
\begin{array}{rrr}
400 & 10 & 13 \\
-100 & -50 & -6 \\
\hline
??? & ??? & 7
\end{array}
$$

Now you have an answer for the ones column. It's 7. But you're left with another problem in the tens column. How do you take 50 from 10? Same story — borrow a single 100 from hundreds column. That reduces the hundreds column by 100, and the problem looks like this:

$$
\begin{array}{rrr}
300 & 110 & 13 \\
-100 & -50 & -6 \\
\hline
??? & 60 & 7
\end{array}
$$

The hundreds column is a no-problem column. Just subtract. The problem now is

$$
\begin{array}{rrr}
300 & 110 & 13 \\
-100 & -50 & -6 \\
\hline
200 & 60 & 7
\end{array}
$$

When you lump the numbers together again, you see the answer is 267. Or to put it another way:

$$
\begin{array}{r}
423 \\
-156 \\
\hline
267
\end{array}
$$

The world's most famous line about borrowing is from Hamlet. Polonius gives some good advice to his son, Laertes, before the boy goes off to college in Paris. Polonius says, "Neither a borrower nor a lender be/ For loan oft loses both itself and friend/ And borrowing dulls the edge of husbandry."

Checking your work

To check subtraction, the rule is simple: Just add the difference back to the subtrahend to get the minuend. It should be the same number you started out to subtract from. For example, in the problem:

$$
\begin{array}{r}
423 \\
-156 \\
\hline
267
\end{array}
$$

add 267 back to 156 (267 + 156 = ???). The answer is 423, the number you started out with.

Example: Flour Power

You have a job at the Berkeley Artisan Bakery, a large commercial bakery (despite its name) with many exotic types of flour in stock, including

Flour	Amount
Almond flour	100 kilograms (kg)
Amaranth flour	50 kilograms
Atta flour	25 kilograms
Bean flour	25 kilograms
Brown rice flour	100 kilograms
Buckwheat flour	250 kilograms
Cassava flour	10 kilograms

How much flour do you have?

This calculation is simple, as are most addition problems.

1. **Form a column with the flour amounts.**

   100 kg
   50 kg
   25 kg
   25 kg
   100 kg
   250 kg
   10 kg

2. **Add the amounts.**

   100
   50
   25
   25
   100
   250
   <u>10</u>
   560

Yo! Nothing to it! The answer is 560 kilograms.

Example: Sheep on Trucking

You're in the Sheep Program, part of the Department of Animal Science, and you're an intern at a ranching operation.

This morning, you counted 1,750 sheep in the pens, and now you have a smaller number. Your supervisor asks you how many trucks picked up the missing sheep today. Can you tell him?

You can't proceed until you confirm two important facts:

- ✔ How many sheep are left in the pens
- ✔ How many sheep a truck holds

Fortunately, the answers are at hand.

- ✔ 850 sheep are left in the pens.
- ✔ Each truck holds exactly 100 sheep.

This setup is, by the way, an example of the classic "double whammy" story problem. The answer isn't about sheep — it's about trucks. But you can get there from here.

Here's how to approach the problem.

1. **Subtract the number of sheep left from the number of sheep you had at the start of the day.**

   The difference is the number of sheep trucked away.

1,750	starting number
−850	remaining number left in pens
900	number trucked away

   900 sheep have been trucked away today.

2. **Figure out how many trucks hauled the sheep away.**

   Do this part in your head. If 900 sheep are gone and each truck holds 100 sheep, how many trucks picked up sheep? If you don't want to do it in your head, do this division.

   $$\frac{900 \text{ sheep}}{100 \text{ sheep/truck}}$$

The answer is nine trucks.

Baa, baa, black sheep

Basque people started coming to America from Spain in the 1850s because of the California gold rush. They became synonymous with "sheepherder" in Nevada. They used black sheep as markers, and the herder would only count the black ones. If they were all with the herd, chances were all of the sheep were together. If a black sheep was missing, the herder and his dogs would set out in search of the missing black sheep and whatever other sheep had gone with this marker.

Chapter 5

Multiplication and Division: Everybody Needs Them

· ·

In This Chapter

▶ Understanding what multiplication and division are and how they work

▶ Performing big-number multiplication without a calculator

▶ Recognizing special cases that make multiplication and division easy

▶ Performing basic short and long division

· ·

*M*ultiplication and division are parts of basic math, and they're as essential as addition and subtraction (see the preceding chapter). The good news is that just about everybody learned multiplication and division in elementary school. The bad news is that many people have forgotten how to do these operations. And if you didn't like math in your earlier school years, you probably forgot even faster.

Note: We're not going to make excuses for you — you need to be responsible for your own knowledge — but if you struggle with math fundamentals, it may not be entirely your fault. 1989 brought a considerable change to math teaching standards, with a decrease in learning fundamentals, so depending on your age, you may have been a victim of the early "new math" instruction and may not have gotten all the basics you needed.

If you had problems with multiplication and division in elementary, middle, or high school, be troubled no more. They're simpler than you may remember.

And they're important. Why? Because they're essential to your work, whether you multiply pounds of cement or bytes in a disk sector or divide fluids or flour. And the conversion of all weights and measures requires multiplication and division. (Check out Chapter 6 for more on measurement and conversion.)

In this chapter, you discover the names of the parts of multiplication and division equations as well as a couple of different ways to multiply and divide.

Go Forth and Multiply!

What do you do if you're at work, attending a meeting, or at an appointment and you need to do some quick multiplication, but your calculator or phone is in your car, which is parked about a zillion miles down the street? You can mumble something about forgetting something and make a mad dash to your car, or you can use the skills in the big carbon-based calculator in your head, your brain. The following sections show you how to take the latter route.

Mastering multiplication terminology

Don't freak out at the thought of unaided multiplication. *Multiplication* is just a form of repeated adding. For example, you typically say the equation $3 \times 4 = 12$ as "three times four equals twelve." But you can also write that equation as $3 + 3 + 3 + 3 = 12$, which you say as "three added four times equals twelve" or "three plus three plus three plus three equals twelve."

And that's repeated adding. This process works pretty well until you have to multiply, say, $459 \times 661 = 303,399$. That's a lot of repeated adding, so in this case you want to use manual multiplication, which we discuss in "Doing Simple Multiplication Like Your Grandfather Did It" later in the chapter.

Every art and craft has its special words, and multiplication is no different. In multiplication, the number to be multiplied is called the *multiplicand,* and the number doing the multiplying is the *multiplier.* The result is the *product.* To make things more interesting (by which we mean "confusing"), sometimes the multiplier and multiplicand are generically called *factors.* Look at the example $459 \times 661 = 303,399$. Here, 459 is the multiplicand, 661 is the multiplier, and 303,399 is the product. The numbers *459* and *661* are factors of 303,399.

 One popular online source says that multiplication was documented in the Egyptian, Greek, Babylonian, Indus valley, and Chinese civilizations. And the Ishango bone, found in the then-Belgian Congo in 1960 and dated to about 18,000 to 20,000 BC, has marks that may suggest knowledge of addition and multiplication. Just speculation about the use, but the marks are real.

You run across a variety of math symbols that represent multiplication. Don't be alarmed. They all mean the same thing: multiply. Sometimes, depending on the problem, you can more easily and cleanly show multiplication by using one symbol rather than another.

Popular symbols include parentheses [( )], a single dot (·) called a *middle dot*, the times symbol (×), or even an asterisk (*). The asterisk is used mainly on computers, adding machines, and on some calculators. Here are these signs in action:

$$(7)(3) \qquad 7 \cdot 3 \qquad 7 \times 3 \qquad 7*3$$

All of these examples are saying "seven times three."

Another sign of multiplication is no sign. This situation occurs in algebra when numbers and letters appear together, such as in the term *3ab*. That means $3 \times a \times b$. We cover algebraic variables more thoroughly in Chapter 12.

Sometimes multiplication is represented as a grid, so you can see the numbers represented as rows and columns. Figure 5-1 shows you how 3×10 can be represented as three rows of ten columns.

Figure 5-1: Multiplication shown as a grid of objects.

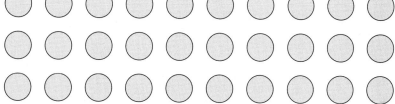

Memorizing multiplication tables: Faster than a calculator

Sources say that the Chinese invented the multiplication table. But regardless of who came up with it, you should commit the multiplication table to memory. It's a must-do thing, and it's not that hard. Here's why nailing the times table is such a worthwhile pursuit:

- You probably learned it in school and may just need to review.
- The multiplication facts involving zero and one are easy (flip to "Easy Street: Multiplying by 0, 1, and 10" later in the chapter for the skinny on these shortcuts).
- It only goes up to 9×9.
- You need it to do longer multiplication and division problems.
- It's faster than a calculator.

Look at that last point. It's true. You can say "seven times seven equals forty-nine" faster than you can punch the numbers into a calculator.

Figure 5-2 shows a classic multiplication table. It's a 9 x 9 but includes 0 as well (so *technically,* it's 10 x 10). You can also find 12 x 12 and 20 x 20 tables.

X	0	1	2	3	4	5	6	7	8	9
0	0	0	0	0	0	0	0	0	0	0
1	0	1	2	3	4	5	6	7	8	9
2	0	2	4	6	8	10	12	14	16	18
3	0	3	5	9	12	15	18	21	24	27
4	0	4	8	12	16	20	24	28	32	36
5	0	5	10	15	20	25	30	35	40	45
6	0	6	12	18	24	30	36	42	48	54
7	0	7	14	21	28	35	42	49	56	63
8	0	8	16	24	32	40	48	56	64	72
9	0	9	18	27	36	45	54	63	72	81

Figure 5-2:
Classic 9 x 9 multiplication table.

To use the multiplication table, find the row you want to multiply (for example, the *3* row). Read across until you come to the column you want to multiply by (such as the *6* column). The answer is where the two meet (in this example, the answer is 18). Figure 5-3 shows how to use the table to find a value.

X	0	1	2	3	4	5	6	7	8	9
0	0	0	0	0	0	0	0	0	0	0
1	0	1	2	3	4	5	6	7	8	9
2	0	2	4	6	8	10	12	14	16	18
3	0	3	5	9	12	15	18	21	24	27
4	0	4	8	12	16	20	24	28	32	36
5	0	5	10	15	20	25	30	35	40	45
6	0	6	12	18	24	30	36	42	48	54
7	0	7	14	21	28	35	42	49	56	63
8	0	8	16	24	32	40	48	56	64	72
9	0	9	18	27	36	45	54	63	72	81

Figure 5-3:
Finding a value in a multiplication table.

Doing Simple Multiplication Like Your Grandfather Did It

Handheld calculators didn't become accessible to businesses and schools until the early 1970s, but of course people had to do multiplication somehow before then. That's where the *traditional multiplication* (the kind you actually do with pencil and paper) in this section comes in. It can serve you when the batteries are dead and also help you gauge the reasonableness of products displayed by a handheld calculator.

It's believed that the original calculator (a kind of abacus) was developed by the Egyptians around 2,000 BC. Other sources say it goes back farther, to ancient Mesopotamia, and others cite China. The first adding machine was developed in the 17th century, and the late 19th century saw the introduction of the first commercially developed adding machine.

Always perform every step of a multiplication problem, and make sure you do each step neatly. If the answer is wrong, you can more easily track the steps you took to arrive at it.

To solve a simple multiplication problem by hand, just follow these steps:

1. **Write out the multiplication problem.**

   Say you want to multiply 23 by 4. Write the factors down as follows, so that the tens and ones columns in each factor line up.

   ```
     23
   × 4
   ```

2. **Multiply the each column in the multiplier by the multiplicand.**

   You multiply 4 times 3 and 4 times 2. The same is true for longer multipliers, no matter how many digits it may have.

   If your multiplication results in a number higher than 9, you record the ones number and *carry over* the tens number to the next column. After you do the multiplication for that next column, you add the carryover number to that result, carrying over again into the next column if necessary.

   ```
    1
     23
   × 4
   ─────
      2
   ```

   The product of 4 times 3 is 12. The result is higher than 10, so just write the *2* in the ones column and carry the *1*.

Then, the product of 4 times 2 is 8. Add the carried *1* and write the result, 9, in the tens column, as shown

```
  1
 23
× 4
────
 92
```

Sometimes, though, you have a multiplier with more than one digit. In those cases, the solving process is a bit more complex.

1. **Follow Steps 1 and 2 of the basic multiplication process earlier in this section.**

   For example, say you want to multiply 7,089,675 by 345. Write the problem down as follows, so that the hundreds, tens, and ones columns in each factor line up.

   ```
     7,089,675
   ×       345
   ───────────
   ```

   Multiply each digit in the multiplicand by the ones-column digit in the multiplier.

   Start with the *5* in the *345* on the bottom. You multiply 5 times 5, 5 times 7, and so on, moving from right to left until you've multiplied by all the digits in the multiplicand. Remember to keep your work organized by bringing each answer straight down, keeping it aligned with its appropriate column.

   The top row in the following example shows the carryovers from this step's multiplication:

   ```
      443,325
    7,089,675
   ×       345
   ───────────
   35,448,375
   ```

2. **Repeat Step 1 with the digit in the tens column of the multiplier, inserting a placeholder of 0 (zero) in the ones column of this multiplying step.**

   You use this zero placeholder because multiplying by 4 is really multiplying by 40 (because the 4 represents 4 tens, or 40, in the context of the whole number). As we note in the following section, any number times 10 shifts its decimal point one place to the right, and in the case of whole numbers, that amounts to adding a zero at the right. So adding this placeholder reminds you to account for the zero in the ones place of 40.

The following shows the placeholder:

443,32

7,089,675

×   345

35,448,375

         0

With the *0* in place you can now multiply the *4* through the multiplicand. Write the result next to the *0* placeholder. The result follows, with the carryovers in the top row:

332 32

7,089,675

×   345

35,448,375

283,587,000

3. **Repeat Step 2 for the remaining digits in the multiplier, adding place-holder zeroes as appropriate.**

As in the previous step (multiplying by a number in the tens column), you need to add placeholder zeroes for the hundreds column, the thousands column, and all other columns in the multiplier. The placeholders keep things lined up and help ensure that the answer will be correct.

The following shows the placeholders for the example problem:

7,089,675

×   345

35,448,375

283,587,000

      00

The following shows the example's complete multiplication for this step, with the carryovers in the top row:

7,089,675

×   345

35,448,375

283,587,000

2,126,902,500

Oh, happy day! The multiplication is now over and just one final step remains.

4. **Add the partial products to get the answer.**

   It's normal addition. Take the first column on the far right and add together 5 + 0 + 0. Continue with each column.

$$
\begin{array}{r}
7{,}089{,}675 \\
\times \quad\quad 345 \\
\hline
35{,}448{,}375 \\
283{,}587{,}000 \\
2{,}126{,}902{,}500 \\
\hline
2{,}445{,}937{,}875
\end{array}
$$

   The answer is 2,445,937,875.

Using a calculator for multiplication and division

These days, your math brain is your pocket calculator and, increasingly, your cellphone. On many phones, you can even find a tip calculator, which is a multiplication application (app). When your mind knows the principles of multiplication (and indeed, other math principles, too), your fingers can work with confidence. Then you can let them do the walking on the calculator or phone keyboard.

Knowing fundamental multiplication and division (including the multiplication table) saves you effort. For example, you don't need waste time with a calculator to multiply 2 × 2.

Doing manual multiplication and division is often faster and easier than using a calculator app. Of course, this knowledge is handy if you don't have your phone with you, but that's not the main reason to know the principles.

Practicing the principles of manual multiplication and division makes you much more aware of the reasonableness of calculator answers. When you have a sense of what a reasonable answer is, you can question calculator results if necessary. This ability is important because a calculator usually produces only a final answer, and it can be way off if you don't pay attention.

Having said that, a calculator is a mighty handy tool. Check out the following example of multiplying two large numbers using a small set of keystrokes:

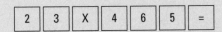

The answer is 10,695. Here's a division example:

The answer is 21.47826. To do this math by hand would take time that you can probably better spend on another task.

Checking your work

To check a product in multiplication, divide by one of the factors. For example, if you do the problem $3 \times 4 = 12$ and you want to check the work, divide 12 by 4 and you get 3, verifying your answer.

$3 \times 4 = 12$

This example is obvious, but the principle is also true for larger numbers, such as this example from earlier in the chapter:

$7,089,675 \times 345 = 2,445,937,875$

Divide the answer by either of the original factors to check your work:

$2,445,937,875 \div 345 = 7,089,675$

$2,445,937,875 \div 7,089,675 = 345$

No matter which factor you choose, the division always yields the other factor as the answer, assuming you've done the math correctly. If you come up with different answer, go back and check your work.

Easy Street: Multiplying by 0, 1, and 10

If traditional multiplication makes you sweat (see the preceding section), remember these three shortcuts that can help you get your multiplication answers more quickly. The shortcuts are simple:

- ✔ Check for multiplication by 0.
- ✔ Check for multiplication by 1.
- ✔ Check for multiplication by 10.

With these shortcuts, you can get your job done quickly, make your boss happy, and enjoy a bright future (well, at least you'll get your job done speedily).

A zero pulse: Multiplying by 0

Anything multiplied by 0 (zero) is just that, zero. No matter what numbers you try to multiply by zero — whole numbers, fractions, decimals — the product is still zero. Find a pattern in the following example.

$$1 \times 0 = 0$$
$$2 \times 0 = 0$$
$$\frac{1}{2} \times 0 = 0$$
$$7\frac{1}{2} \times 0 = 0$$
$$-29 \times 0 = 0$$

It's obvious. But if you don't believe us yet, look at any multiplication table (such as the one in Figure 5-2 earlier in the chapter): The 0 column and 0 row both produce 0 as the product in every one of their squares. When you see "$\times 0$" in a math problem, just call the answer 0 and move on.

One is the loneliest number: Multiplying by 1

Actually, one may or may not be the loneliest number, but for simplicity, it's in a strong competition with its brother 0. Anything multiplied by 1 is itself. Period. The following examples illustrate the point:

$$2 \times 1 = 2$$
$$14 \times 1 = 14$$
$$23 \times 1 = 23$$
$$6,099 \times 1 = 6,099$$
$$\frac{1}{2} \times 1 = \frac{1}{2}$$
$$0.87323 \times 1 = 0.87323$$

Knowing this rule helps you get to the answer and move on to a more productive task. What about zero? Well, when it's multiplied by 1, the answer is itself, 0 ($0 \times 1 = 0$). See the preceding section for more on multiplying by zero.

Multiplying by 10

If you're multiplying a multiplicand by 10, a great shortcut is to just put a zero to the right of the multiplicand and stop. That's the answer.

For example, the product of 3 and 10 is probably obvious to you: $3 \times 10 = 30$.

Notice that the product is the multiplicand, 3, with a 0 attached.

If you do the math manually, you see that this shortcut works because the decimal number system is a positional system (see Chapter 3) and because of the principles you apply to exponents of 10 (see Chapter 11). That is, multiplication by 10 is the same as shifting a number's decimal point one place to the right.

You can extend the principle to other multiples of ten:

- ✔ If you're multiplying a multiplicand by 100, put 00 to the right of the multiplicand.
- ✔ If you're multiplying a multiplicand by 1,000, put 000 to the right of the multiplicand.

What about 5×20 (which equals 100)? That's not multiplying by 10, but you know that 20 is really 2×10. So put the ideas together by separating 20 into 2 and 10.

$$5 \times 20 = 5 \times 2 \times 10$$

There! You have just factored 20 into 2 and 10. Multiply 5 by 2, giving 10, and then add the *0* (for multiplying by 10). That produces 100. And you did it in your head.

Divide and Conquer

You probably remember a collective sigh (or groan) in the third grade when the teacher told you that you were going to do division. It's traditionally the most dreaded basic math operation out there. And if you didn't get it in the third grade, they threw it at you twice as hard in the fourth grade.

Division is the opposite (or *inverse*) of multiplication. Multiplication is repeated adding, so division is repeated subtracting. For example, consider $12 \div 3 = 4$.

You say this equation as "twelve divided by three equals four." Here is an example of getting the answer by repeated subtraction:

$$12 - 3 - 3 - 3 - 3 = 0$$

If you subtract 3 from 12 four times, you have nothing left to subtract. This method works, but it's pretty tedious and certainly wouldn't be fun with big numbers.

The following sections introduce you to the division dictionary and help you get the hang of the various kinds of division. For real-world problems, your calculator will likely be on hand, but knowing the techniques here can help you out in a calculator-less pinch and improve your understanding of what you're doing with the calculator. We also give you some shortcuts to relieve the division headache just a bit.

Dealing with division definitions

Like multiplication, division has special names for its components. The number you divide into is called the *dividend,* and the number you divide by is called the *divisor.* The result is the *quotient.* Take a look at the following example:

101,439 ÷ 221 = 459

In this example, 101,439 is the dividend, 221 is the divisor, and 459 is the quotient.

What if your division doesn't come out in a nice, even quotient? The *remainder* is what's left over when the dividend can't be evenly divided by the divisor. In the example 13 ÷ 3, the divisor 3 goes into 13 four times, but that just makes 12. You have a remainder of 1, the part that can't be evenly divided. You write this as r1 after the four (4 r1).

To be very formal about it, the algorithm for division can be stated as $a = bq + r$, with qualifiers as to number type, range, zero and non-zero conditions, and remainder conditions. It's very impressive, but it's not good reading.

Division has a few different math signs, but they all amount to the same thing — time to divide. They're largely interchangeable, but sometimes choosing the right division sign makes your work easier. You may see the classic division symbol (÷) or a forward slash (/), which is the division sign of choice on a computer keyboard. Sometimes division appears as a *stacked fraction,* or an *inline fraction* like the one found in the third of the following examples.

$$7 \div 3 \quad \frac{7}{3} \quad \frac{7}{3}$$

All of these examples mean "seven divided by three."

One special tool for division is the *tableau,* from the French for "table" or "picture." When it's empty, it looks like this:

$$\overline{)}$$

When it has numbers in it, it looks like this:

$$3\overline{)12}^{\,4}$$

The dividend goes inside the tableau. The divisor is at the left. The quotient (as you develop it) goes above.

Dividing by using the inverse

Because division is the inverse of multiplication, one way to divide is to *invert* the divisor (divide it into 1) and *multiply*. **Note:** Using the inverse isn't the common method, but we do use it in our fraction discussion in Chapter 8, so we want to tell you about it here.

Say you want to divide 35 by 7. Normally, the expression is 35 ÷ 7 = ?.

Get the inverse by dividing the divisor into one. The inverse of 7 is

$$\frac{1}{7}$$

So for the task of dividing 35 by 7, you can set up this *multiplication* problem:

$$35 \times \frac{1}{7} = ?$$

Now, just calculate the answer, which is 5.

Doing short division

Short division is a simple, fast way to do division. It just takes a pencil and paper, and you can do a lot of it in your head. Knowing your multiplication table makes it faster (and luckily, we present one in Figure 5-2 earlier in the chapter). Basically, short division breaks the dividend into chunks.

But short division has limitations. It works better with smaller divisors, up to about 12, and if the problem gets complicated, you've got to switch to long division.

A simple case of short division

Here's a simple division problem, with no remainders.

Imagine you have a high volume of sulfuric acid — 48,488 fluid ounces, to be exact. You have four carboys to store them in. How many ounces do you put in each carboy?

Use the following short division process to find the answer:

1. **Write the dividend and divisor in the tableau.**

   In this case, you're dividing 48,488 by 4, so your setup looks like the following:

   $$4\overline{)48,488}$$

2. **Divide the divisor into the first digit or digits larger than the divisor and write the quotient above the digit.**

   If the divisor is too big to go into only the first digit, try the first two digits. For this example, the digit must be 4 or greater or it won't fit. If the first digit were 3, the divisor (4) wouldn't fit, and you'd divide into the first two digits.

   $$\begin{array}{r} 1 \\ 4\overline{)48,488} \end{array}$$

3. **Repeat Step 2 for the remaining digits.**

   Here's what that looks like for the second digit and the completed problem:

   $$\begin{array}{r} 12 \\ 4\overline{)48,488} \end{array}$$

   $$\begin{array}{r} 12,122 \\ 4\overline{)48,488} \end{array}$$

   That's it! The answer is 12,111 ounces per carboy.

Simple short division turns more complex

Sometimes, your short division doesn't come out as neatly and remainder-free as you may want. Check out the following twist on the example from the preceding section to see how you can use short division to tackle more complicated problems.

You have four carboys to store 972 ounces of sulfuric acid in. You follow much the same process as the problem in the preceding section does, but you have to account for remainders.

1. **Divide the divisor into the first digit where it fits.**

   The first digit is 9. The divisor 4 goes into 9 two times, but with 1 left over. Write a *2* above the 9 in the quotient area, and put a little 1 to the right of the nine. That's the remainder.

   $$\begin{array}{r} 2 \\ 4\overline{)9^{1}72} \end{array}$$

2. **Divide into the second digit, using any remainder from the previous step as a tens digit.**

The remainder of 1 and the second digit of 7 make 17. Four goes into 17 four times ($4 \times 4 = 16$) with a new remainder of 1, which you record to the right of the second digit.

$$4\overline{)9^17^12}$$
$$2\ 4$$

3. **Repeat Step 2 for the remaining digits.**

The remainder of 1 and the third digit of 2 make 12.

$$4\overline{)9^17^12}$$
$$2\ 4\ 3$$

In this case, your answer has come out even, so you're done. The answer is 243.

If the result doesn't come out even, you have a remainder, which you note next to the quotient. For example, if you divide 4 into 973, the result is 243 with a remainder of 1.

$$4\overline{)9^17^12}$$
$$2^14^13\,r1$$

Going long (division)

Short division (see the preceding sections) is useful for simpler problems, and a calculator is helpful for really complex (or "hairy," to use a professional term) problems.

But in the middle is long division, which some claim is the path to spiritual fulfillment. Real men and women know how to do long division. They have confidence and more fulfilling lives (or so we hear). Long division is excellent for long numbers, both in the dividends and/or the divisors, because it breaks the problem down into very clear steps.

Keep your work neat and organized. It makes a difference, especially if you have to go back and dissect the work to find an error.

Follow these steps to perform long division:

1. **Write the dividend and divisor in the tableau.**

Say you want to tackle the problem 24,432 ÷ 16. For long division, write the problem like this:

$$16\overline{)24,432}$$

2. **Divide the divisor into the first digits where it fits.**

If the divisor is too big to go into only the first digit, try the first two digits. In this example, 16 doesn't fit into 2, the first digit, but it does fit into 24, the first two digits. Write the quotient above the digits, placing it over the last digit you went into.

$$
\begin{array}{r}
1 \\
16\overline{)24{,}432}
\end{array}
$$

3. **Multiply the quotient from Step 2 by the divisor, write the product below the first two digits of the dividend, and then subtract.**

   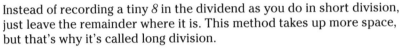

   Here, 1 times 16 is 16, so record that below the *24* in the dividend. 24 minus 16 gives you a remainder of 8.

   Instead of recording a tiny *8* in the dividend as you do in short division, just leave the remainder where it is. This method takes up more space, but that's why it's called long division.

$$
\begin{array}{r}
1 \\
16\overline{)24{,}432} \\
16 \\
\hline
8
\end{array}
$$

4. **Bring down the next digit from the dividend and place it next to the remainder from Step 3; divide the divisor into this new number, writing the answer in the quotient area.**

   The next number in the dividend is 4, so pull that down next to the remainder 8 from Step 3 to make 84. Divide 84 by 16 and put the answer, 5, in the quotient, above the digit you dropped down. You now have 15 in the quotient area.

$$
\begin{array}{r}
1\,5 \\
16\overline{)24{,}432} \\
16 \\
\hline
84
\end{array}
$$

5. **Repeat Steps 3 and 4 until you can't divide into the dividend any more.**

   Here's what your final setup for this example looks like:

$$
\begin{array}{r}
1{,}527 \\
16\overline{)24{,}432} \\
16 \\
\hline
84 \\
80 \\
\hline
43 \\
32 \\
\hline
112 \\
112 \\
\hline
0
\end{array}
$$

   The answer is 1,527.

This example is a simple division problem, with no remainders. If your problem doesn't come out evenly, just note the remainder next to the quotient. If you need a more precise answer, simply extend the dividend with a decimal point and one or more zeroes and keep on dividing.

Smaller dividends take less time. Larger dividends take more time. If you start dividing numbers like 5,439,242 by numbers like 649, pull out the calculator. (But do be careful. Some calculators can only handle numbers so big and may give you an error message if the numbers are too large.

Checking your work

To check a quotient in division (where else would you find quotients?), multiply the quotient by the divisor; if you've solved the problem correctly, this check returns the dividend.

For example, if you do the problem $12 \div 3 = 4$ and you want to check the work, multiply 4, the quotient, by 3, the divisor. The result is 12, your original dividend. To check the problem from the long division section earlier in the chapter, multiply 16 by 1,527:

$$16 \times 1,527 = 24,432$$

Shortcuts: Dividing into 0 and by 0, 1, 10, and the dividend

Like multiplication, division offers a few scenarios that can make your life (or at least your math) a little easier. Check your problem for

- A dividend of 0
- A divisor of 0
- A divisor of 1 or of the dividend itself
- A divisor of 10

Dividing into 0

Zero divided by anything is zero. Stop working. That's the answer.

Dividing by 0

The rule is simple: Never divide by zero. Never. You can't. Dividing by zero is a mathematical impossibility. It's a waste of time, and it annoys the zero.

Dividing by 1 or the dividend

As with multiplication, dividing by 1 is simple. Any number divided by 1 is equal to (get ready) itself. Stop working. That's the answer. Similarly, any number divided by itself is 1.

Dividing by 10

If you're dividing 10 into a dividend, just move the decimal point one place to the left. For example, $5,450 \div 10 = 545$.

This concept works for dividing by 100 as well. Divide by 100 by moving the decimal point two places to the left: $67,400 \div 100 = 674$.

Example: In the Machine Shop

Old lathes can be driven by belts. If a pulley makes 4,522 revolutions in 34 minutes, how many revolutions does it make in 1 minute? You know how many revolutions the pulley makes in 34 minutes, so just divide that number by 34. Set it up as a long division problem.

1. **Put the numbers from the problem into the tableau.**

   Your dividend is 4,522, and your divisor is 34.

   $34\overline{)4{,}522}$

2. **Divide the divisor (34) into the first digits where it fits.**

   It doesn't fit into 4, the first digit, but it does fit into the first two digits, 45. Write the quotient above the digits.

   $\dfrac{1}{34\overline{)4{,}522}}$

3. **Multiply the quotient (1) by 34 and write it below the first two digits (45); then subtract.**

   $$\begin{array}{r} 1 \\ 34\overline{)4{,}522} \\ \underline{34} \\ 11 \end{array}$$

4. **Bring down the next digit (2); divide 34 into 112 and write the answer above the tableau.**

You now have 13 in the quotient area.

$$\begin{array}{r} 13 \\ 34\overline{)4{,}522} \\ \underline{34} \\ 112 \end{array}$$

5. **Multiply the quotient (3) by 34 and write it below the 112; subtract (leaving a remainder of 10) and bring down the next digit (2).**

Now you have a divisor of 102.

$$\begin{array}{r} 13 \\ 34\overline{)4{,}522} \\ \underline{34} \\ 112 \\ \underline{102} \\ 102 \end{array}$$

6. **Divide 34 into 102 and write the result (3) in the quotient area; subtract.**

You have no remainder, so you're done.

$$\begin{array}{r} 133 \\ 34\overline{)4{,}522} \\ \underline{34} \\ 112 \\ \underline{102} \\ 102 \\ \underline{102} \\ 0 \end{array}$$

The answer is 133 revolutions per minute.

Chapter 6

Measurement and Conversion

In This Chapter
▶ Identifying measurement systems
▶ Converting from one kind of unit to another

*I*n the early history of civilization, measuring things was probably a key building block helping people move from a nomadic life to one based on farming and living in villages. Common questions then no doubt included things like: "How long is it?" "How much does it weigh?" "What is its volume?" and "What is its area?"

Today people still need answers to the same questions. For example, carpenters, roofers, masons, cabinet makers, and painters use measurements of length (feet and inches) extensively. Chefs and cosmetologists use volume measurements (fluid ounces and cups). Lab assistants use weights and volumes in the metric system (kilograms and liters).

The units for measuring things have helped civilization from its start. They have also plagued each civilization when it traded with the civilization next door and the units of measurement didn't agree. This problem still exists today because two major systems are in use. Some of these units may be plaguing you now. But not to worry. Having two measurement systems isn't a problem when you see how it all works.

In this chapter, you identify the different systems of measurement and see the most common units. Then you review how to convert from one unit to another, whether the units are in the same or different systems.

Main (And Not So Main) Systems of Measurement

Today you commonly use two systems of measurement: the American system and the metric system. The American system is also called the

United States customary system. The metric system is now known as the International System of Units (abbreviated *SI*), though we refer to it here as the metric system.

These aren't the only systems of measurement, however. Many countries all over the world have official systems of measurement, some of which go back thousands of years. These are *customary* systems, like the United States customary system, but those systems have been almost universally overtaken by metric.

Some minor systems of measurement are important, but only in a very limited number of careers.

In some cases, your work requires knowledge of only one system. For example, science and medicine use the metric system almost exclusively, while the construction trades and culinary arts use the American system extensively. Stay alert, though: The American system is a tradition, but metric measurement is gradually becoming more widely used in the United States.

- ✔ In photography, you use millimeters (mm) to specify lens focal length.

- ✔ In heating/cooling (HVAC), the Joule is likely to replace the British thermal unit (BTU).

- ✔ In nutrition, the kilojoule (kJ) will probably replace the kilocalorie (kcal or Cal).

- ✔ In cooking, you want to get familiar with kilograms and liters if you want to cook in the United Kingdom or Europe.

The U.S. National Institute of Standards and Technology, part of the U.S. Department of Commerce, says, "In keeping with the Omnibus Trade and Competitiveness Act of 1988, the ultimate objective is to make the International System of Units the primary measurement system used in the United States." Here's a new word to impress your friends: The process of introducing the metric system to the United States is called *metrication*.

The metric system

The official name may be the International System of Units (SI), but you commonly refer to it as the *metric system.* It's based on a unit of length called the *meter.*

The metric system is a *decimalized* system. That means each larger unit of length, area, volume, or mass (weight) is ten times the size of the previous smaller unit. For example, there are names for 1 meter, 10 meters, 100 meters, and 1,000 meters: meter, decameter, hectometer, and kilometer. You know the last one — it's the same kilometer you see on road signs.

In 1791, the French Academy of Sciences defined the meter (or *metre,* as much of the world spells it) as one ten-millionth of the distance from the equator to the north pole through Paris. In 1983, the General Conference on Weights and Measures redefined the meter as the length of the path travelled by light in vacuum during a time interval of

$$\frac{1}{299,792,458}$$

of a second. You'll surely sleep better at night for knowing.

So whether you prefer *meter* or *metre,* it's still the same thing. It's just a little more than a yard long (1 meter ≈ 39.37 inches — see the following section). That little squiggle (≈) means *approximately.* That's about the height of a parking meter on a city street, but that has nothing to do how the meter was named. Sources say the word is from Greek, meaning *a measure.*

For a long time, the metric temperature scale was called the *centigrade* scale — with water freezing at 0 degrees centigrade and boiling at 100. Then they named it after Anders Celsius, a Swedish astronomer. But this system didn't account for absolute zero — the lowest temperature theoretically possible — so now 0 kelvin (0 K) is defined as –273.15 degrees Celsius, or absolute 0. However, when you drive by a bank with an electric sign, you see degrees Fahrenheit (the American temperature unit) and degrees Celsius, but not kelvin.

In time measurement, a *second* is the duration of 9,192,631,770 periods of the radiation corresponding to the transition between the two hyperfine levels of the ground state of the caesium 133 atom.

The big advantage in using metric units (aside from most of the world using them) is that most of the conversions are multiples of 10. Now is that easy math, or what?

The American system

The United States is the only industrialized nation that doesn't use metric for standard activities. Burma (Myanmar), Liberia, and the U.S. are the only countries that haven't adopted metric. Instead, the U.S. uses the *American system.*

The American system is sometimes called the *English system,* or its units are called *English units.* It dates back to colonial days. In the United Kingdom, on the other hand, the English system evolved to become the imperial system (see the section on the imperial system later in this chapter). The American system is similar, but not identical, to the imperial system.

Table 6-1 shows you how American units roughly compare to metric units. They aren't exact matches, but you can see which American units have similar metric counterparts.

Table 6-1   Comparing Main American and Main Metric Units

American	Metric
Length	
	millimeter
inch	centimeter (2.54 inches)
foot (12 inches)	
yard (3 feet)	meter (39.37 inches)
rod (16.5 feet)	
furlong ($\frac{1}{8}$ mile)	
mile (5,280 feet)	kilometer (0.6214 miles)
Area	
square inch	square centimeter (0.1550 square inches)
square foot (144 square inches)	square decimeter (0.1076 square feet)
square yard (9 square feet)	
acre (43,560 square feet)	hectare (10,000 square meters, 0.01 square kilometers, 0.4047 acres)
square mile (640 acres)	square kilometer (0.3861 square miles)
Volume	
	cubic millimeter
cubic inch	cubic centimeter (0.0610 cubic inches)
cubic foot (1,728 cubic inches)	cubic meter (0.0283 cubic feet)
cubic yard (27 cubic feet)	cubic meter (0.7645 cubic yards)
Weight	
	gram
grain	
dram (27.3438 grains)	
ounce (437.5 grains)	gram (28.35 grams)
pound (16 ounces, 7,000 grains)	gram (454 grams)
	kilogram (1,000 grams)
hundredweight (100 pounds)	
ton (20 hundred weight, 2,000 pounds)	tonne (1,000 kilograms, 1.1023 tons)
long hundredweight (112 pounds)	
long ton (20 long hundredweight, 2,240 pounds)	

American	Metric
Liquid volume	
	milliliter
fluid ounce (29.5735 milliliters)	
cup (8 fluid ounces)	milliliters (237 milliters)
pint (16 fluid ounces)	
	liter (1,000 milliliters)
quart (32 fluid ounces, 2 pints)	liter (1.0567 quarts)
gallon (4 quarts)	
Temperature	
degrees Fahrenheit	degrees kelvin

The American pound is also known as *avoirdupois* weight, as opposed to the *troy* pound (see "Troy weight: Just for bullets and bullion") or the *apothecaries'* pound (see "Apothecaries' system: Not a grain of value any more" in this chapter). The key to distinguishing the American weight unit is that avoirdupois has 16 ounces in 1 pound. The others don't.

For information on how to convert between the metric and American systems, see "Converting Length, Weight, and Volume" later in this chapter.

The imperial system, or the modern English system

The imperial system and imperial units are still common in the United Kingdom, which transitioned to the metric system legally in 1995 but still has many imperial units in use.

The U.K. Department for Business, Innovation and Skills decrees that draft beer and cider must be sold in pints or half pints. (However, gin, rum, whiskey, and vodka should be in sold in multiples of 25 or 35 milliliters and glasses of wine in 125 milliliter portions.)

Road sign distances and speed limits have to be expressed as imperial units. You find similar situations in Ireland and Canada, where people are permitted or even required to use some imperial units. You also find some imperial units in use in former British territories: Australia, India, Malaysia, New Zealand, South Africa, and Hong Kong.

Just remember that many *but not all* American units are identical to imperial units.

Troy weight: Just for bullets and bullion

Troy weight is a system for measuring weight (mass). It's still very much alive, but it's used mainly for weighting precious metals and reloading rifle and pistol cartridges. Its units are

- Grain
- Pennyweight
- Ounce troy
- Pound troy

The unit grain is the same in avoirdupois, troy, and apothecaries' weight.

Propellant (powder) for firearms is still weighed in grains but is sold by the avoirdupois pound. Bullets are identified by their weight in grains but are sold in packages with a specific quantity.

Unlike the American avoirdupois pound, which has 16 ounces, troy weight has 12 ounces to its pound. It's a smaller pound than an avoirdupois pound.

A *carat* is a unit of weight for a gemstone. It is equivalent to 200 milligrams (mg) or $\frac{1}{5}$ gram (g). Also, in the world of jewelry, you can divide a carat into 100 points. So, if someone gives you a 5-point diamond, there has to be a lot of love, because there isn't much diamond.

Don't get carat confused with karat (k), which is a measure of the purity of gold or silver, as in 14-karat gold.

Apothecaries' system: Not a grain of value any more

Apothecaries' weight is very similar to troy weight (the grain and the pound are the same, with each kind of pound containing 5,760 grains) and is rarely used. It is an old system for apothacaries (pharmacists) and scientists.

The units of the apothecaries' system are:

- Grain
- Scruple
- Dram
- Ounce apothecaries
- Pound apothecaries

What is a grain, anyway?

Way back when, traders used wheat or barley grains to define weight. They also used the carob seed (which is where you get carat weight of gemstones). In the 13th century, King Henry III of England declared that an English penny was to be equal to 32 grains of wheat. The grain is the same in avoirdupois, troy, and apothecaries' weight.

Apothecaries' grains are still used occasionally in medical prescriptions. Although most dosages are in milligrams (mg), some items may still be sold in grains (gr).

A pound apothecaries has 12 ounces. It's a smaller pound than an avoirdupois (American) pound, which has 16 ounces, but the same weight as a pound troy.

Other legitimate but specialized measurements

As long as there are many things to measure, there will be many units of measurement. Over the centuries, systems developed as the trades or science required them.

Some of these systems seem to be passing into obsolescence, but as the world changes and technology evolves, new systems come into existence:

- ✔ **Writing paper:** In the world of the printing press, the following units evolved:

  - Sheet

  - Quire

  - Ream

  - Bundle

  - Bale

  You often buy printer paper at the office supply store by the ream. If you buy a case of ten reams, you've also bought a bale.

- ✔ **Light-years and parsecs:** In the world of astronomy, you measure large distances, such as the distances to stars. A light-year is the distance that light travels in a year. And since light travels very fast (299,792,458 meters per second), that's a big distance. It's about 5,878,630,000,000 miles or exactly 9,460,730,472,580.8 kilometers. A parsec is a unit of length equal to about 3.26 light-years. The star nearest the earth (not counting the sun) is 1.29 parsecs away.

✔ **Wine containers:** As time marches on, some units do not march on. For example, English units of wine cask capacity have pretty much disappeared, so bid farewell to the rundlet, the tierce, the hogshead, the firkin, the puncheon, the tertian, the pipe, the butt, and the tun. They were once common units but have been overtaken by changing times.

✔ **Board feet:** Carpenters in the United States and Canada commonly use the board foot to order lumber. BF is the most common abbreviation, but some say the abbreviation is FBM, for "foot, board measure." A *board foot* is a volume measurement equal to a 1-inch-thick board 12 inches wide and 1 foot long. That's 1 "linear foot" of a 1 x 12. You also express that as 12 inches x 12 inches x 1 inch. It amounts to 144 cubic inches.

Even though 1 x 12 boards are no longer 1 inch thick or 12 inches wide, those dimensions are still used in calculating board feet. The calculation uses the *nominal* width and thickness. The unit applies to other lumber as well. For example, an 8-foot-long 2 x 4 contains 2 x 4 x 96 cubic inches, or 768 in^3, or 5.333 board feet.

✔ **Pressure:** Pressure is the force applied to a unit of area. The metric unit for pressure is the pascal (Pa), formerly Newton per square meter (N/m^2). In the American system, the unit for pressure is pounds per square inch (psi). Pressure measurement is important when working with water supplies, welding gases, and medical gases.

✔ **Flow:** Flow is the measurement of a fluid. That usually means a liquid, but *fluid* can also mean air or other gases. You measure flow with flowmeters. It's important when working with water supplies, including drilling water wells. For well drillers, gallons per minute (gpm) is the most critical and common unit of flow used.

✔ **Electricity:** The symbol for power in equations is *P,* but electricians use the unit *watt,* which is shorthand for *electrical power in watts.* It's the rate at which energy is generated and consumed. Residential, commercial, and industrial electricians use units such the watt (W), the ampere (A, also known as the amp), and the volt (V) every day.

Power electricians install and maintain power generators, converters, transformers, and distribution networks. Electricians are multitalented constructors. They read blueprints, install tubes (conduit) in the walls, install boxes to hold switches and outlets, and pull the wires. At the same time, electricians knows how to do the math to deliver the expected power to each part of the structure, and the safety requirements for keeping the installation safe.

Converting Length, Weight, and Volume

In some careers, you probably can't do a day's work without converting a number from one unit to another. For example, a framing carpenter constantly converts feet to inches and inches to feet in order to make cuts. A concrete pourer calculates square feet for a slab's area, calculates cubic feet of the pour, and then converts to cubic yards to know how much concrete is required.

Converting units isn't hard. You just need to know the tricks and then practice a little bit.

The rules of conversion

To convert from one unit to another, keep three things in mind:

- ✔ Know the formula.
- ✔ Use identical units.
- ✔ Use the right conversion tools.

Know the formula

You can almost always find a formula on the Internet or in a book to convert from one unit to another. However, many formulas are so simple that you probably have them memorized.

- ✔ A foot has 12 inches.
- ✔ A yard has 3 feet.
- ✔ A mile has 5,280 feet.

If you grew up with the American system, you probably find this conversion especially easy, even though the units aren't so easy.

It's only slightly more of a challenge to convert from American to metric or metric to American, but between tables of conversion, formulas, and your memory, you should have no problem.

Use identical units: Don't add feet to furlongs!

When you see math with different units, that's a sign that you need to do some conversion. You can't directly add inches to miles or feet to furlongs. You need to convert to the same units before doing the math.

Sometimes the dimensions or quantities are accurate but very inconvenient. You need to convert. For example, here's a small patio that needs a concrete pour:

3 yards × 108 inches = ? square feet

This multiplication isn't going to work. You can't multiply yards by inches and get square feet. The formula for the area of a rectangle, which you probably remember from elementary school, is

$a = l \times w$

Length times width equals area. But the units have to be the same. For the patio, that formula should be:

a (in square feet) = l (in feet) × w (in feet)

The units (feet) are the same, and the result (square feet) will be correct. You know that 1 yard is 3 feet long, so convert 3 yards to get 9 feet You know that 1 foot contains 12 inches, so divide 108 inches by 12 to get 9 feet. The following shows the new equation, now that you've converted 3 yards and 108 inches:

9 feet × 9 feet = ? square feet

Now you can easily multiply to get the answer in square feet.

Use the right conversion tools

Conversion is faster and more accurate if you use the right tools, which may be tables, a calculator, pencil and paper, or the math stored in your brain. Here's the story of Goldilocks and the Three Conversion Problems:

✔ The first problem was too complex. Goldilocks had to combine small metric volumes of reagents with larger American volumes of water to create three jugs of solution for a home urine test. Goldilocks used a table of conversions and a calculator.

✔ The second problem was too simple. Goldilocks had to cut a 3-foot piece of plywood into 9-inch strips. She did the math in her head, converting 3 feet into 36 inches and dividing by 9. She cut four strips.

✔ The third problem was just right. Goldilocks had to combine several fluid ingredients in a lab. They were measured in milliliters and liters. She knew how to convert (1,000 milliliters = 1 liter), but she wrote the items down and added them manually.

American units to American units

When you work with American units, you encounter many different unusual equivalents. However, converting from one unit to another isn't hard to do.

Whether the conversion involves a relatively simple number (for example, 1 yard is equal to 3 feet) or a more complex-looking number (for example 1 mile is equal to 5,280 feet), you use conversion factors to get the job done.

Length and distance

The general way to do any conversion is as follows:

Given Unit × Conversion Factor = Desired Unit

The number of inches in a foot is a basic conversion factor:

1 foot = 12 inches

To do a conversion, write the conversion factor as a fraction. You can write it two ways. They are equivalent and are both correct.

$$\frac{1 \text{ foot}}{12 \text{ inches}} \qquad \frac{12 \text{ inches}}{1 \text{ foot}}$$

Use the first fraction

$$\frac{1 \text{ foot}}{12 \text{ inches}}$$

to convert inches to feet. To convert 36 inches to feet:

$$\frac{36 \text{ inches}}{1} \times \frac{1 \text{ foot}}{12 \text{ inches}} = \frac{36 \text{ feet}}{12} = 3 \text{ feet}$$

Notice how the inches in the top and bottom cancel each other out. You are left with feet. This conversion is actually division; it's just shown as multiplying by the inverse (check out Chapter 8 for more about working with fractions).

You can also do the math this way:

36 inches ÷ 12 inches in a foot = 3 feet

Use the second fraction

$$\frac{12 \text{ inches}}{1 \text{ foot}}$$

to convert feet to inches. To convert 3 feet to inches:

$$\frac{3 \text{ feet}}{1} \times \frac{12 \text{ inches}}{1 \text{ foot}} = \frac{36 \text{ inches}}{1} = 36 \text{ inches}$$

Notice how the feet in the top and bottom cancel each other out. You are left with inches. This conversion is multiplication.

If you get outrageous results, that's a check that tells you that you multiplied or divided the wrong way.

Weight (mass)

Converting units of weight is identical to converting units of length.

Given Unit × Conversion Factor = Desired Unit

One of the most common weight conversions is from ounces to pounds. Another common conversion is the *inverse*, where you convert from pounds to ounces.

16 ounces = 1 pound   1 pound = 16 ounces

The scale in a doctor's office usually weighs adults to within about a quarter of a pound. A typical weight for an adult woman is

$$136\frac{1}{2} \text{ pounds}$$

No conversion is required. But if you measure the weight of an infant as 142 ounces, you need to convert.

To convert this figure to pounds and ounces, apply the conversion factor 16 ounces = 1 pound.

$$\frac{142 \text{ ounces}}{1} \times \frac{1 \text{ pound}}{16 \text{ ounces}} = \frac{142}{16} \text{ pounds}$$

That's mathematically correct, but you can't tell mom and dad that the baby weighs

$$\frac{142}{16}$$

pounds. The calculator result is 8.875 pounds, which is mathematically correct, but again, that's not the easiest result for the parents to understand.

So continue the conversion. Divide 142 by 16, but leave a remainder:

$$\frac{142}{16} \text{ pounds} = 8 \text{ pounds, with a remainder of } \frac{14}{16} \text{ pounds}$$

The answer is 8 pounds, with

$$\frac{14}{16}$$

of a pound remaining. That's 14 ounces, so the plain English answer is that baby weighs 8 pounds, 14 ounces.

If you work in a doctor's office, you may use a body mass index (BMI) meter to get a sense of whether a patient is obese or not. The meter does the work for you. However, it does ask for the patient's height and weight. Depending how you've set up the meter, you can enter American units or metric units. You can also find BMI calculators on the Internet.

What is *mass,* anyway? Aren't you talking about weight? Yes and no. Put on your Full Baffle Protector: *Mass* is an intrinsic property of matter, whereas weight is a force that results from the action of gravity. That doesn't mean much on Earth. A mass of 100 pounds has a weight of 100 pounds. On the moon, however, the same mass weighs 16.666 pounds (since the moon's gravity is about $\frac{1}{6}$ of Earth's gravity). If you plan to stay on Earth, you can treat mass and weight as the same thing.

Volume

If you've read the preceding sections, it's probably occurred to you that converting units of volume is identical to converting units of length and weight.

Given Unit × Conversion Factor = Desired Unit

For example, if you know these conversion factors:

2 pints = 1 quart   4 quarts = 1 gallon

you can calculate the number of pints in a gallon. This conversion isn't trivial when you are creating institutional meals for hospitals and school food services.

$$\frac{2 \text{ pints}}{1 \text{ quart}} \times \frac{4 \text{ quarts}}{1 \text{ gallon}} = \frac{8 \text{ pints}}{1 \text{ gallon}}$$

The answer is 8 pints per gallon.

But wait! There are 16 fluid ounces in a pint. How many fluid ounces are in a gallon?

$$\frac{16 \text{ fluid ounces}}{1 \text{ pint}} \times \frac{2 \text{ pints}}{1 \text{ quart}} \times \frac{4 \text{ quarts}}{1 \text{ gallon}} = \frac{128 \text{ fluid ounces}}{1 \text{ gallon}}$$

Just put in another conversion factor, $\frac{16 \text{ fluid ounces}}{1 \text{ pint}}$. The answer is that there are 128 fluid ounces in a gallon.

American to metric and back again

Metric units appear more often and in more places. From time to time, you need to convert from U.S. customary units to metric or from metric to U.S. customary units.

You get a little informal help with this just by looking at common occurrences in everyday life. Dietary information on food is shown in metric. A 12-ounce can of soda also has *355 mL* printed on it. Even measuring cups in your kitchen have both U.S. and metric units on them.

Length and distance

The conversion factor for feet to meters is

1 foot = 0.3048 meters

As an example, a person 6 feet tall is 1.83 meters tall.

Inches are sometimes more useful than feet. The conversion factor for meters to is

0.0254 meters = 1 inch

As an example, an 84-inch doorframe is 2.13 meters high.

You may want to convert miles to kilometers or kilometers to miles when you drive. The conversion factor for miles to kilometers is

1 mile = 1.609344 kilometers

The conversion factor for kilometers to miles is

1 kilometer = 0.6213712 miles

Weight (mass)

How much do you weigh in kilograms? The conversion factor for pounds to kilograms is

1 pound = 0.454 kilograms

The conversion factor for kilograms to is

1 kilogram = 2.2 pounds

Volume

In the U.S., gasoline is sold by the gallon, but in many countries it's sold by the liter.

The conversion factor for gallons to liters is

1 gallon = 3.785 liters

The conversion factor for liters to gallons is

1 liter = 0.264 gallons

Converting metric to metric

Metric-to-metric conversions are probably the easiest ones to make. Add one or more zeroes to multiply. Remove one or more zeroes to divide. And every unit is a multiple of another unit.

Length and distance

How many meters are in a kilometer? The answer is in the definition: The prefix *kilo* is from Greek, and it means *1,000.*

1,000 meters = 1 kilometer

The kilometer isn't used widely yet in the United States, but it's appearing on road signs, next to the mile numbers. The United States and United Kingdom are the only developed nations that still use miles on road signs.

You can say *kill*-o-meter or kill-*om*-etter. They're both okay. The second pronunciation is consistent with barometer, thermometer, tachometer, and speedometer.

Weight (mass)

The simple answer about weight conversions in metric is the same as the answer for length and distance. The answer is in the definition.

How many grams are in a kilogram?

1,000 grams = 1 kilogram

With medications, the milligram (mg) and the microgram (mcg) are common.

1,000 milligrams = 1 gram

1,000 micrograms = 1 milligram

The prefix *milli-* indicates that a gram contains 1,000 milligrams. Unlike *kilo,* it's not a multiple. Because it's smaller than the base unit, it's called a *submultiple.*

Volume

Please don't enter a persistent vegetative state. There is one more metric truth to share. How many milliliters are in a liter? The answer is — again — in the definition:

1,000 milliliters = 1 liter

Example: Don't Get Bored by Board Feet

You work for a fencing contractor. The boss asks you to order 240 8-foot 4 x 4 fence posts. How many board feet is that? You know that a board foot is 144 cubic inches.

1 board foot = 1 inch × 12 inches × 12 inches = 144 cubic inches

1. **Convert the 8-foot dimension to inches, so all units for the 4 x 4 are in inches.**

$$8 \text{ feet} \times \frac{12 \text{ inches}}{1 \text{ foot}} = 96 \text{ inches}$$

2. **Multiply thickness, width, and length to get the volume of a single 4 x 4.**

4 inches × 4 inches × 96 inches = 1,536 cubic inches

3. **Multiply by the number of 4 x 4s you need to buy.**

1,536 cubic inches × 240 = 368,640 cubic inches

4. **Divide by the conversion factor (1 board foot = 144 cubic inches) to get the result in board feet.**

The result is the total volume in cubic inches.

$$368,640 \text{ cubic inches} \times \frac{1 \text{ board foot}}{144 \text{ cubic inches}} = 2,560 \text{ board feet}$$

The answer is 2,560 board feet. Whether the lumberyard sells in board feet or *per each,* that's going to be a serious investment in fence posts.

Example: Getting the Dosage Right

This example is drawn from actual nursing experience. You are to give "gr 5 FeSO4 [iron sulfate]" but the bottle contains 325-milligram tablets of iron sulfate. How many milligrams is the order for?

You look up grains in an apothacaries' weight table and find that 1 gram equals 15.432 grains.

1. **Convert the 5 grains into grams.**

$$5 \text{ grains} \times \frac{1 \text{ gram}}{15.432 \text{ grains}} = 0.324 \text{ grams}$$

2. **Convert the grams into milligrams.**

A milligram is

$$\frac{1}{1,000}$$

of a gram:

$$0.324 \text{ grams} \times \frac{1,000 \text{ milligrams}}{1 \text{ gram}} = 324 \text{ milligrams}$$

The answer is 324 milligrams, which is almost exactly equal to the 325-milligram tablets in the bottle.

Chapter 7

Slaying the Story Problem Dragon

You encounter story problems every day. They appear in two major places: math tests and your work. You can avoid story problems in math tests by completing all the math classes you need for an education in your career, but you can't avoid story problems that arise in your everyday work.

Very few problems come with a formula attached; instead, they come with words attached. For example, if you have a recipe for one pie and want to enough pies to serve 100 people, it's up to you to calculate the number of servings in one pie to, figure out the number of pies you need and then convert the recipe into the right amounts of ingredients. It doesn't matter whether you're the boss or just the apprentice baker — the problem will lie there until you solve it.

Story problems are disconcerting to people, making them break out in a cold sweat, make groaning noises, and sometimes curl up into a ball. We researched this effect, and we find almost 100-percent agreement. People seem to fear and loathe story problems.

In the past, you may have heard story problems called *word problems* or *life problems,* and the reaction was worse. Don't hyperventilate. Story problems are "real-life" problems, and they aren't hard to solve. You can slay the biggest fire-breathing story problem dragon if you have a big enough fire extinguisher to douse the flames. Then it just takes a penknife (or a ballpoint pen) to finish it off.

In this chapter, you look at what's inside a story problem and discover a series of steps (such as finding the keywords) that make just about any story problem solvable.

Removing the Mystery from Story Problems

Story problems come in many flavors. The classic ones are money or investment problems, distance/speed/time problems, and age problems ("How old will John be when he's half Mary's age?").

The point (and the good news) is that most classic types of story problems aren't important to your work. In the trades, you're far more likely to encounter area, perimeter, or volume problems; combining or separating mixture problems; and/or number or weight problems.

Identify the types of story problems you're faced with and practice solving those types. In yoga, martial arts, and story problems, practice is a component of success. The more story problems you solve successfully, the greater your confidence in solving new ones.

The promise of a career with limited types of story problems doesn't guarantee an education with limited types. While preparing for any career, vocation, or trade, you may encounter math courses with story problems that don't apply to your work — anything from problems about "Betty's age" to "two trains leave a station, going in opposite directions." Stay alert and you won't get hurt!

The illusion of a magician pulling a rabbit out of a hat is mystifying — until you know the trick. The same is true with story problems. They appear confusing, until you know the trick of how to approach them. Lucky for you, we discuss that topic in the following section.

How to approach a story problem: A real-life example

Here's an irony of life. Story problems on a math test give you all the information you need but are designed to contain some mystery. Story problems in real life aren't supposed to be mysterious, but sometimes they don't have all the information. Here's an example:

You're a skilled fencing installer. You know materials (wood and steel), concrete, and fabrication techniques. Here's a sample problem:

> How many six-foot-high chain-link fence posts spaced ten feet apart do you need to fence a one-acre enclosure?

This example is just about the "perfect storm" of story problems. It has a little of everything good or bad you may find in a story problem.

Look for two major symptoms:

- ✔ **Too much information:** When a story problem has too much information, identify that fluff and eliminate it. For example, in the fencing example, the object is to figure the number of posts around an enclosure. The fact that the posts are six feet high is interesting but not important to solving the problem.

  As we advise in the step-by-step solving method later in this chapter, be sure to read the problem carefully so that you can identify what information is pertinent to your needs.

- ✔ **Too little information:** When a story problem has too little information, identify what's missing and get what missing information you need. Sometimes you can convert the information provided, and sometimes you have to ask your source for it. In the fencing example, you calculate the number of fence posts needed to surround an acre. But the problem doesn't tell you how big around an acre is.

To get off to a brave start, look up the area of an acre: 43,560 square feet. To restate the question, what's the perimeter ("how big around") of an acre? The answer: You don't know. You can't know. Even the area info you looked up isn't terribly helpful — any number of dimensions can yield that figure. You may have an enclosure 43,560 feet long and one foot wide. That's a whole lotta fence posts. In real life, you have to go back to the source for more information.

The smallest rectangle that can enclose an acre is a square with sides of 208.71 feet. The fence posts in the example would have to be enough to handle 834.84 feet of perimeter. A circular enclosure would take fewer fence posts because it has a perimeter (well, circumference) of 739.86 feet.

An American football field is 1.322 acres in area. (Don't count the end zones.) One acre is 75.6 percent of the football field's area.

Say the fencing enclosure isn't an acre but rather a rectangle with dimensions of 50 feet x 100 feet. The perimeter (the total distance around all four sides) is 300 feet, so if the fence posts go in the ground every 10 feet, you need 30 posts. (Flip to Chapters 15 and 16 for more on perimeters.)

What if it's not an enclosure? What if it's a straight run of 100 feet? The answer is 10 fence posts, right? Wrong! This scenario is where experience counts. You must make allowance for one more post at the start of the run, which brings your total to 11 fence posts.

What if the run (whether straight or an enclosure) isn't evenly divisible by 10 feet? What if it's 94 feet rather than 100 feet? As a practical matter, you can't just install the posts 10 feet apart; the fence will look bad at the end of the run. To get the cosmetics right, a professional consults a fence post spacing chart. The spacing for a 94 feet run should be 9 feet, 5 inches. Also, as a pro, you know what corner posts, end posts, and gate posts are, when you need them, and how they affect your figures. Others have to look that information up.

In a math test, the problem doesn't change because of these kinds of considerations. In real life, real life can change the problem.

The secret formula inside every story problem

A lot of story problems finally turn into a simple math problems. The initial facts may garble the message at first, but when you work them out, the math message becomes clear.

Chapter 13 in this book contains detail about formulas. But for now, just think of a *formula* as a rule or principle usually expressed in symbols, not words. For example, the formula for the area of a rectangle ("area is equal to length multiplied by width") is $A = L \times W$.

Secret formulas aren't so secret, except in the movies and on TV. Math formulas are more common, and they work all the time. A story problem converts into a formula. In fact, you may be a little disappointed when you see the formula and the mystery disappears. The following sections show you how to dig the formula out of a story problem.

Recognizing the real-life factors

Life has a funny way (funny peculiar, not funny har har) of being more complex than made-up problems. (Well, you may make an exception for afternoon soap operas.) So although math test story problems are designed to test one or maybe two math concepts, real-life story problems require you to work with multiple math principles.

The answer to one part of a story problem usually leads to one or more story problems within the story problem. These *multipart story problems* require you to solve the main part of the problem and then solve other parts on top of that. Some people feel like they're being stabbed with a knife covered with poison. Is the multipart story problem really deadly? No, not really. The "deadly" part is that there's simply a second (or third) solving step after the first one.

Money and labor issues frequently find their way into otherwise simple problems. If you're a fence installer, your math may not end with figuring the number of fence posts needed for a fencing run. You expect it to go on to calculate money or labor, and you encounter these factors:

- ✔ The cost of special posts
- ✔ The height of the posts (to determine the height of the fencing fabric)
- ✔ The total cost of the fence and the average cost per running foot
- ✔ The hours needed for installation
- ✔ The number of installers and the overall installation time

Two-part story problems may also ask you to do one of two kinds of conversions:

- ✔ **Direct conversion:** Different portions of the problem are given in different units. For example, you usually calculate a volume of concrete to be poured based on feet, giving you cubic feet, but the problem may ask you to specify how much concrete to order, which is based on cubic yards. You usually convert the cubic feet to cubic yards.

- ✔ **Indirect conversion:** The initial answer is based on one unit but the final answer isn't a simple conversion to another unit. For example, in tiling a roof, if you know the area of a roof, you have to convert the roof area in square feet area into the number of roof tiles that cover the area. But roof tiles overlap, so you do a conversion to allow for overlap. Then you convert to order the tiles by count, not area. The final answer isn't the first or second intermediate answer.

Identifying the keywords

Another mystery in story problems comes from the words, which you have to turn into math expressions. The idea is to translate from words to math. Table 7-1 shows some typical keywords and their probable meanings.

Table 7-1	Typical Keywords and Their Probable Meanings
When you see these words	*They probably mean*
Combined, total, sum, added to, increased by	Addition
Difference, fewer than, less than, decreased by, left over, remaining	Subtraction
Times, multiplied by	Multiplication
Per, percent, out of	Division

Table 7-2 shows some other keywords — in this case, units that suggest mathematical operations.

Table 7-2	Units that Suggest Mathematical Operations
When you see these words	*They probably mean*
Specific units (such as apples and oranges) to be combined into a general unit (such as pieces of fruit)	Addition
A general unit (such as pieces of fruit) to be separated into specific units (such as apples and oranges)	Subtraction
Number of items with a small area (such as shingles) needed to cover a large area (such as a roof)	Multiplication
An item with a large area (such as a roof) needing a number of items with a small area (such as shingles) needed to cover it	Division

The Step-by-Step Story Problem Solution

If you can't bear the thought of tackling story problems, just think of them as puzzles. All you need to conquer a puzzle is a method for solving it, and the method for tackling this particular kind is in this section.

To succeed with story problems, keep your thoughts and work organized.

The steps in the following sections show you how to clear up story problem mysteries in an orderly step-by-step way. *Note:* We use an example of making a big batch of chocolate mousse to illustrate many of the steps, but we offer other examples when they show the point of the step better.

1. Read the problem

To redo an old saying, "Don't just do something. Stand there!" Don't calculate anything at this point. Begin by reading the problem carefully. It's simple and it's true: You must read the whole problem. Don't make any marks. Then reread the problem so that you understand what answer you need and what information you have (steps we cover in later sections).

Here is a classic example of a (nonmath) problem that people don't fully read before coming up with an answer:

> **Question:** A plane crashes exactly on the border between two states. Where do you bury the survivors?

> **Answer:** Half of them in each state? All of them in one state? In a long thin line along the border? The answer: nowhere. You don't bury survivors. In this example, your brain gets ahead of your eyes, and you come up with an answer (the wrong one) too soon.

2. List the facts

Identify and list the facts. Look at all of the information given in the story problem and make a list of what you know. If the problem requests an answer (and what problem doesn't?), you determine exactly what it is in a different step. If any of the info is irrelevant, you determine that in a different step.

Pretend that you're an assistant manager in the food service at an upscale retirement complex. You serve a lot of meals to a large number of diners. You must determine the amounts of the ingredients you need for chocolate mousse for 250 guests, based on the following recipe notes:

The ingredients for one batch are as follows:

- ✔ 2 cups heavy cream
- ✔ 4 large egg yolks
- ✔ 3 tablespoons sugar
- ✔ 1 teaspoon vanilla
- ✔ 7 ounces fine-quality bittersweet chocolate

Nutrition facts: Total fat 32.3 grams, saturated fat 18.5 grams, and cholesterol 283 milligrams. The mousse goes into eight 6-ounce stemmed glasses and is garnished with lightly sweetened whipped cream.

The facts are easy to list:

- ✔ Need 250 servings
- ✔ 2 cups heavy cream
- ✔ 4 large egg yolks

✔ 3 tablespoons sugar

✔ 1 teaspoon vanilla

✔ 7 ounces fine-quality bittersweet chocolate

✔ Lightly sweetened whipped cream (for garnish)

✔ Nutrition elements are total fat 32.3 grams, saturated fat 18.5 grams, and cholesterol 283 milligrams

✔ Eight 6-ounce stemmed glasses

3. Figure out exactly what the problem is asking for

Know what you're trying to find. The problem often states the required answer, but sometimes you have to ferret it out from the information you receive. In the recipe example in the preceding section, the answer is clear: You have to determine the amount of ingredients required for 250 servings.

4. Eliminate excess information

A problem may have excess information — in fact, some math test problems may include extra facts on purpose. For example, you may need to find out how many oranges are in Jim's basket, but the problem may also tell you that Jim is wearing a yellow shirt. Too much information! The trouble is that in real-life problems, identifying the excess information may be harder.

In the earlier recipe example, eliminate the following irrelevant facts:

✔ Lightly sweetened whipped cream (for garnish)

✔ Nutrition elements are total fat 32.3 grams, saturated fat 18.5 grams, and cholesterol 283 milligrams

✔ 6-ounce stemmed glasses

Why are these items irrelevant?

✔ The sweetened whipped cream garnish isn't important to the main problem, calculating amount of ingredients for the mousse. The recipe doesn't even specify the necessary amount of whipped cream, so you can't calculate the amount you'd need anyway.

✔ The nutritional elements are outrageous, but not relevant. Nutrition has no bearing on solving the main problem.

✔ The number of servings in the batch is important, but the size of the serving glasses isn't. You may not fill the full six ounces of each glass, so the serving may be any number of ounces.

Important information may be lurking near the junk you don't need, so be vigilant. If you weren't careful here, you may have thrown out the fact that you're using *eight* 6-ounce glasses; the ounces may be unimportant, but the eight servings per batch is key to figuring out how many batches serve 250.

5. See what information is missing

Determine what information is missing. In a test problem, you can usually convert some piece of given information into information you need but don't have. In real life, you may also have to seek out more information.

For example, in commercial housepainting, you easily calculate the square footage of walls to be painted, and you also know the number of square feet a gallon of paint can cover. But if you don't know whether to apply one coat or two, you calculation of total amount of paint needed may be significantly wrong. You can't do the calculation successfully until you know the missing piece of information. In the case of housepainting, you need more information about the covering ability of the paint to be used and about the color to be painted over.

In the earlier recipe example, the information is complete. You don't need to find more info. You don't need to convert units either, but beware: Later, as a practical matter, you convert the answers into units that are familiar to your buyers.

6. Find the keywords

As we note in the earlier section "Identifying the keywords," story problems often contain words that you need to translate into nice, useful math symbols and formulas. (Check out Tables 7-1 and 7-2 in that section for some handy translations.) In the recipe problem, the keywords are "determine the amounts of ingredients" and "for 250 guests."

This first phrase tells you that the answer requires some new amount of ingredients. This process is called *scaling* a recipe and asks you to find either greater or lesser quantities of ingredients. Scaling asks for addition or multiplication when you're increasing a recipe and subtraction or division when you're decreasing a recipe.

The second phrase tells you that the quantities of ingredients must be correct for 250 servings. Since the original recipe serves 8, the phrase implies that you must multiply to get the amounts needed.

7. Pay attention to units

Units make a difference. If an answer requires the results in feet, that's how the answer must appear. Unfortunately, problems sometimes use units that aren't the same as those required for the answer. When that happens, be prepared to convert units. You can use a table of conversions, an online calculator, or your own memory to get the conversion factors.

For example, if you have a carpet remnant 72 inches wide and 108 inches long, you can easily multiply length and width to get the area: 7,776 square inches. However, when was the last time you saw carpet measured in square inches? Carpets typically come in square feet, so you need to convert the answer.

A square foot contains 144 square inches (because it's a measure of area 12 inches in length and 12 inches in width), so divide 7,776 by 144 to get 54 square feet. You can also get this answer by converting 72 inches into 6 feet and 108 inches into 9 feet. When you multiply 6 and 9, you get 54 square feet.

When the units used in calculating are different from each other, you convert them to the same unit. For example, you can't add feet to meters. Convert the feet to meters or the meters to feet so that you're doing math with the same unit. See Chapter 6 for more on converting units.

8. Convert information supplied into information needed

In a perfect world, you get the units you need, but alas, the world isn't perfect. In everyday work, you get mixed units, whether the items involved are quantities of fence posts, containers of ethyl alcohol, or drums of transmission fluid.

Be prepared to convert the elements of the problem into the units you need. Why say "elements?" We're not talking strictly about units of measure. You may also be changing the form of the substances in the problem.

For instance, in a problem about preparing peach compote for a large food service, your starting recipe may call for cups of sliced peaches. However, your produce supplier may deliver whole fresh peaches by the pound or lug. Use your professional skills to determine how many cups of peach slices you get from a pound or a lug.

The *lug* is a unit used in agriculture. It's a box used to hold fruits or vegetables. The United States Department of Agriculture defines lugs for peaches as 22 pounds.

9. Draw a diagram

A diagram helps you visualize the problem. For example, a straight 100-foot run of chain link fencing with posts set 10 feet apart suggest that you need 10 fence posts, as we discuss earlier in the chapter. But if you draw a diagram (such as the one in Figure 7-1) to verify, you discover that you actually need 11 posts to cover 100 feet at 10-foot intervals.

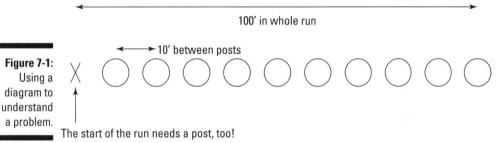

100' in whole run

Figure 7-1:
Using a
diagram to
understand
a problem.

10' between posts

The start of the run needs a post, too!

Diagrams are frequently helpful, but they're not useful for every problem, so think of this step as an option, not a requirement.

10. Find or develop a formula

Formulas for converting units (for example, feet to inches or feet to meters) are common and necessary. A formula assures you that your reasoning is right and may adapt to become the formula you use for solving the problem. Although existing boilerplate formulas (such as the one for area) may be helpful, what your story problem may need is a formula that deals in that particular problem's units or items.

In the earlier mousse recipe example, you want to take the quantities you have for eight servings and convert them into the quantities you need for 250 servings. The following formulas help you do just that:

$$\frac{\text{quantity for 8 servings}}{8} = \text{quantity for 1 serving}$$

$$\text{quantity for 1 serving} \times 250 = \text{quantity for 250 servings}$$

You can generalize these formulas to apply to any conversion of servings.

11. Consult a reference

If you're stuck, look for a reference of some kind, such as a conversion chart or even a blog where someone has encountered the same problem you're having.

For example, you may want to use a recipe from the United Kingdom and you convert metric units into traditional American units. If you are visiting the United Kingdom and cooking for friends, you convert your recipe with traditional units into metric units.

Look at the following recipe for Indian naan. It has both traditional and metric units:

- 175 milliliters (6 fluid ounces) warm water (45 degrees Celsius)
- 1 teaspoon dried active baking yeast
- 1 teaspoon caster sugar
- 250 grams (9 ounces) plain flour
- 1 teaspoon salt
- 4 tablespoons *ghee* (butter with the milk solids removed)
- 2 tablespoons plain yogurt
- 2 teaspoons onion seed

This recipe happens to include a conversion from milliliters to fluid ounces and from grams to ounces. The other ingredients are in the familiar units of teaspoons and tablespoons.

You can also use your experience as a reference. Experienced chefs get a sense of exactly how much butter, sugar, and whipping cream a recipe requires — they can visualize the quantities from experience. For example, after you know that a quarter-pound stick of butter (4 ounces) is equal to 113.5 grams, you quickly visualize how much butter recipes requiring 90 grams are calling for.

12. Do the math and check your answer to see whether it's reasonable

After you have an answer, be sure to check it. If the result is outrageously high or low, you made a mistake. In the first recipe example earlier, if the chocolate quantity comes out to be 1,750 ounces (over 109 pounds!), you probably need to redo the solution. You didn't divide by 8 to get the quantity for a single serving. The correct answer for 250 servings is 218.75 ounces, or about 13.67 pounds.

Example: Furring Strips

As a remodeler, you know that *furring strips* are 1 inch x 2 inch thin strips of wood you use to make a backing surface for a wall covering (for example, when you drywall a concrete or concrete block basement). You also know that for garage shelves, 2 inch x 2 inch lumber makes a great support for the back edge of plywood shelves.

You're doing a residential remodel and you get a load of lumber. The load has 942 pieces but only two items: 1 x 2 furring strips and 2 x 2 boards. Furring strips cost $1.25 each and 2 x 2 boards cost $2.50 each. The total bill for the load is $1,230.00. How many pieces of each kind of wood do you have? Use the process in "The Step-by-Step Story Problem Solution" to figure it out:

1. **Read and reread the problem.**

2. **Identify things that you know.**

   - Total pieces: 942

   - Total bill: $1,230.00

   - Cost of single furring strip: $1.25

   - Cost of a single 2 x 2 board: $2.50

3. **Determine what the problem is asking you to find.**

   The problem clearly states that you want to know how many pieces of each kind of lumber are in the load.

4. **Eliminate any irrelevant information.**

   It's a residential remodeling project, but that's irrelevant to solving the problem.

5. **Identify keywords.**

   - The problem says, "How many pieces of each kind of wood do you have?" Because you know the total number of pieces (942), you can get the number of the other kind of wood by subtracting when you know the number of pieces of one kind of wood.

   - The problem says, "The total bill for the load is $1,230.00." You add the cost of each kind of lumber together to get the total cost; because you know the cost of each kind of wood, you can use those numbers to help find the amounts of each kind.

6. **Make a formula to solve the problem.**

   Find the cost of the 2 x 2 boards first. You know that each one costs $2.50, so write the formula for the total cost of the 2 x 2s as $x \times 2.50$. You're using a variable because you don't know the exact number of boards yet.

Find the cost of the furring strips. You don't need a new variable; you know the total number of boards delivered (942), so just subtract the number of 2 x 2 boards (which you already have a variable for: x) from 942 to represent the number of furring strips. You can express the number of furring strips as $942 - x$.

That last part took skill and cleverness. You have expressed the number of furring strips in terms of the number of 2 x 2s. Because you know the total bill comes from the total amount of lumber, the formula looks like this:

$$1,230.00 = x \times 2.50 + (942 - x) \times 1.25$$

7. Do the math to solve the problem.

$$1,230.00 = x \times 2.50 + (942 - x) \times 1.25$$
$$1,230.00 = 2.5x + (942 \times 1.25 - 1.25x)$$
$$1,230.00 = 2.5x + 1,177.50 - 1.25x$$

Subtract 1,177.50 from both sides of the equation.

$$1,230.00 - 1,177.50 = 2.5x - 1.25x$$
$$52.50 = 1.25x$$
$$\frac{52.50}{1.25} = \frac{1.25}{1.25}x$$
$$42 = x \text{ This is the number of } 2 \times 2s$$
$$(942 - x) = 900 \text{ This is the number of furring strips}$$

The load contains 900 furring strips and 42 2 x 2s.

8. Check your work.

Plug your answers in and check the results:

$$(42 \times \$2.50) + (900 \times \$1.25) = \$1,230.00$$

These numbers of boards produce the correct costs to equal the total bill.

Example: And Now, from the Banks of the Nile

Talk about the curse of the Pharaohs! Story problems go back to *at least* the Second Intermediate Period of Egypt. This problem comes from a scribe named Ahmes writing in about 1600 BC. Cue the spooky, exotic music:

There are seven houses; in each house there are seven cats; each cat kills seven mice; each mouse has eaten seven grains of barley; each grain would have produced seven *hekat* (an Egyptian unit of grain volume). What is the sum of all the enumerated things?

No surprises in this problem: Reading and examining the problem is straightforward.

1. **Count the houses.**

   The problem itself tells you there are seven houses, so that's easy.

2. **Determine the number of cats.**

   If seven houses each have seven cats, multiply to get the number of cats.

   $7 \times 7 = 49$ cats

3. **Compute the number of mice, grains, and hekats.**

   If each cat kills seven mice, multiply to get the number of mice. Do the same for grains and hekats.

   $7 \times 7 \times 7 = 343$ mice

   $7 \times 7 \times 7 \times 7 = 2,401$ grains

   $7 \times 7 \times 7 \times 7 \times 7 = 16,807$ hekats

4. **Add up the houses, cats, mice, grains, and hekats.**

   Add $7 + 49 + 343 + 2,401 + 16,807$. The answer is 19,607 items.

Part II

Making Non-Basic Math Simple and Easy

The 5th Wave   By Rich Tennant

The fugitives left 200 miles from here traveling 70 miles per hour. We're traveling in the opposite direction going 60 miles per hour. At what point ... hey, donut shop, donut shop! Pull over!

In this part . . .

Arithmetic with whole numbers is fine, but the world runs on more than whole numbers. In Part II, you find out every way to slice and dice a number.

Chapters 8 through 10 all deal with ways to work with a part of 1 (fractions, decimals, and percentages, respectively). Chapter 11 works with exponents, which help you express huge and tiny numbers in a more compact manner, and square roots, which help you find basic dimensions when you know areas.

Chapter 8

Fun with Fractions

. . . .

. . . .

You use fractions without even thinking. You even do the math in your head. You can say "I'll meet you at a quarter past three." That's a fraction. Or you tell a friend, "I started with a full tank of gas, but now I only have 3/4 of a tank. I must have used 1/4 tank this afternoon." That's using fractions and doing math with them.

Fractions are common in technical work. Carpenters, drywallers, chefs, landscapers, cosmetologists, roofers, pastry makers, and concrete pourers (to name but a few) make measurements all the time, and most of those measurements include fractions.

The framing carpenter tells the apprentice to "make sure those studs are 92⅝ inches long." In the culinary arts (mainly cooking and pastry making), measuring and converting fractions are essential. The professional hair colorist uses fractions when mixing 10 Volume or 20 Volume peroxide developer with different proportions of distilled water to make a 5 volume solution for "refreshing" color or doing ends.

Yet some folks get sweaty palms and a chill down the spine when it comes to using fractions in math. No need for that! Fractions are friends — commonplace, easy to work with, essential for getting work done, and fun to use (well, maybe that's stretching it). To succeed with fractions, you just need to know some fraction terms and practice some fraction math.

In this chapter, you find out the names of the parts of a fraction and some names for different kinds of fractions. Then you go on to do simple (but very important) fraction math.

Meeting the Numerator and Denominator: Best Friends Forever

A *fraction* is a number that represents parts of a whole. A fraction is also a way of showing division (we get into that more in Chapter 9). The following numbers are all fractions:

$$\frac{1}{3} \quad \frac{1}{4} \quad \frac{1}{2} \quad \frac{3}{4} \quad \frac{7}{8}$$

A fraction is made up of two numbers and a line. Sometimes a fraction is written like this:

$$\frac{1}{2}$$

and sometimes it's written like this:

$$\frac{1}{2}$$

Here's a chance to impress your friends! If the line is horizontal, it's called a *viniculum*. If the line's slanting (a forward slash), it's called a *solidus*.

The top number (or the left number in the second example) is called the *numerator*. The bottom number (or the right number in the second example) is called the *denominator*. A fraction has to have both a numerator and a denominator. They would be lost without each other. They're best friends forever. Together, the numerator and the denominator are called the *terms* of the fraction.

By the way, *fraction* comes from the Latin verb *frangere*, meaning *to break*. It's usually a small number "broken off" from a whole number. It's also related to *fracture*, as with a broken bone.

How fractions appear in this chapter

Many of the fractions in this chapter appear as *stacked* fractions. A stacked fraction looks like this:

$$\frac{1}{7}$$

It is written straight up and down.

In your everyday activities, you often see fractions written *unstacked* or *inline*. They look like this:

1/7

The inline representation of a fraction might be convenient for casual writing and some reports, but it's not the best way to represent fractions in math operations.

Taking a look at numerators

The numerator can be any number, even a fraction. Think of these numbers as though you were cutting up a cake. (It's a piece of cake.)

For example:

✔ 7/8 represents a cake with 8 pieces, and you have 7 of them.

$$\frac{7}{8}$$

In Figure 8-1, the shaded portion is 7/8 of the cake.

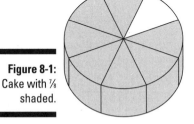

Figure 8-1:
Cake with ⅞ shaded.

✔ 0/8 represents a cake with 8 pieces, and you have none of them. A fraction with 0 in the numerator is a legitimate fraction.

$$\frac{0}{8}$$

In Figure 8-2, the shaded portion is 0/8 of the cake. No shading! That's nothing. You have zero, nada, nil, cipher, goose egg, bupkis.

✔ ½/8 represents a cake with 8 pieces, and you have ½ of one piece.

$$\frac{\frac{1}{2}}{8}$$

In Figure 8-3, the shaded portion is ½/8 of the cake. This must be what dieting looks like.

✔ 8/8 represents a cake with 8 pieces, and you have all 8 of them.

$$\frac{8}{8}$$

In Figure 8-4, the shaded portion is 8/8 of the cake. That's the whole cake, and you're very lucky — or very greedy.

✔ 1/1 represents a cake with 1 piece, and you have the only piece. As with 8/8, you have the whole cake.

$$\frac{1}{1}$$

Figure 8-2:
Cake with
⁰∕₈ shaded.

Figure 8-3:
Cake with
½/8 shaded.

In Figure 8-5, the shaded portion is ⅟₁ of the cake. Now you can tell people, "I only took one piece."

✔ 15/8 is a little trickier. It represents a cake with 8 pieces per cake and you have 15 pieces. That means the 15 pieces come from more than one cake. At first, this may not make a lot of sense, unless you're a caterer with 15 guests to feed.

$$\frac{15}{8}$$

In Figure 8-6, the shaded portion is 15/8 of the cake. If no one asks for seconds, this setup will serve all 15 guests, with one piece left over.

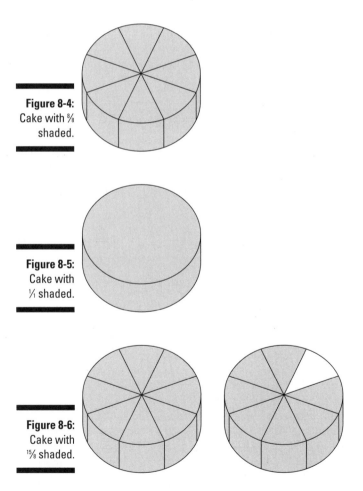

Figure 8-4:
Cake with ⅞ shaded.

Figure 8-5:
Cake with ⅟₁ shaded.

Figure 8-6:
Cake with ¹⁵⁄₈ shaded.

Defining denominators

The denominator can be any number, even a fraction, but with one big exception. The denominator can *never* be 0 (zero). You can have one-fifth of something or one-third of something, but you can't have one-zeroeth of anything. The top number can be 0, but the bottom number can't because it makes no sense, mathematically. Look at these examples of denominators:

- In the fraction 7/8, 8 is the denominator.

$$\frac{7}{8}$$

- Lab techs may use the fraction 50/1,000 to indicate part of a liter. It represents 50 milliliters.

$$\frac{50}{1,000}$$

- Machinists can use 6/1,000 as another way of expressing 6 mils (since a mil is 1/1,000 of an inch).

$$\frac{6}{1,000}$$

- The following example shows a denominator of ½. Fractions in the denominator (as well as the numerator) are allowed, but usually you convert the whole fraction to another fraction that's easier to work with.

$$\frac{154}{\frac{1}{2}}$$

Numerators can be anything — but unlike numerators, denominators can never be 0.

Dealing with special cases

A denominator of 0 can be called a special case, because it messes up the math and isn't possible. You come across several other special cases in fraction math, though, that aren't toxic.

Three cases — 1 as the denominator, 0 as the numerator, and the same number as both numerator and denominator — are special, because you use them all the time to solve math with fractions in the fastest and easiest way. The fourth case — fractions representing cents in a dollar — is special because transactions involving fractions of dollars are common in business and personal finance.

✔ **When 1 is the denominator:** Any fraction where 1 is the bottom number (for example, ³⁴⁄₁) is equal to the top number. The 1 is known as the *invisible denominator*. A math rule: Any number divided by 1 is equal to that number.

$$\frac{34}{1} = 34$$

✔ **When 0 is the numerator:** Any fraction with 0 as the top number (for example, ⁰⁄₁₇) is equal to zero. If you have 0 parts of something made of 17 parts, you have 0 parts. A math rule: When any number is divided into 0, the result is 0.

$$\frac{0}{17} = 0$$

✔ **When the numerator and the denominator are the same, as in 1/1, 3/3, 12/12, and so on:** A math rule: When any number is divided into itself, the result is 1. (These fractions are called *equivalent* fractions, and you find out more about them later in this chapter.)

$$\frac{1}{1} = \frac{3}{3} = \frac{12}{12} = 1$$

✔ **Dollars and cents:** You know that 100 cents make up one U.S. dollar (and about sixty other countries in the world also divide their currency into cents). You also know that a cent (a penny) is 1/100 of a dollar. If you have 37 cents in your pocket, you have 37/100 of a dollar. Currency can be expressed as a fraction where the denominator (100) never changes.

$$\frac{37}{100} \text{ dollars} = 37 \text{ cents}$$

Tackling the Different Types of Fractions

Knowing the names of the different types of fractions is a big help to you in this chapter and when doing math in the real world. You even use one of them in everyday life. In the following sections, we discuss three concepts in detail:

✔ **Proper and improper fractions:** Fractions have two important and obvious forms: A *proper fraction's* denominator is larger than the numerator, and an *improper fraction's* denominator is smaller than the numerator. ("Honey, does this denominator make my fraction look big?")

✔ **Mixed numbers:** Whole numbers are from Mars. Fractions are from Venus. A *mixed number* is what you get when you combine them.

✔ **Ratios:** A ratio is a fraction, sort of, much like your Uncle Willie is a relative until he misbehaves at Thanksgiving dinner. A *ratio* shows two numerators that add up to be a denominator, which may make them deceiving.

Does the preceding list cover every type of fraction? Nope! As proof that mathematicians don't get out much, you may (or may not) want to know that there are also unit fractions, dyadic fractions, Egyptian fractions, continued fractions, and partial fractions. Some of these are special cases (unit and dyadic), some are archaic (Egyptian), and some are used only in higher mathematics (continued and partial).

Proper and improper fractions

Proper and *improper* have nothing to do with good or bad behavior. When a fraction has a numerator smaller than its denominator (smaller number on top), it's a *proper* fraction.

$$\frac{5}{9}$$

When a fraction has a denominator smaller than its numerator (larger number on top), it's an *improper* fraction.

$$\frac{9}{5}$$

So what's so important about this? When you encounter fractions in your work, they will often be part of a mixed number (see the following section). You usually convert the mixed number into an improper fraction so that your calculations are faster and easier to do. Then after doing the math, you convert the result back into a mixed number. This occurs a lot with carpenters, electricians, and workers laying flooring or carpeting.

Mixed numbers

When a whole number and a fraction appear together, such as in

$$2\frac{7}{8}$$

it's called a *mixed number*.

The number represents 2 whole units and 7/8 of another unit.

In various occupations, you find mixed numbers used a lot.

- ✔ A carpenter may cut a piece of plywood to be 24¾ inches x 48 inches.

- ✔ A carpetlayer may trim a roll of carpet to fit in a room that is 111½ inches wide.

- ✔ An electrician may cut EMT (conduit) to a length of 27⅝ inches.

A mixed fraction example

You have a piece of electrical metal tubing that is 8 feet, 9 inches long. How long is that in inches?

The length 8 feet, 9 inches may not appear to be a mixed fraction, but it is. The 8 feet is the number of whole feet; the 9 inches is a fraction of a foot. There are 12 inches in a foot, so 9 inches is 9/12 of a foot.

To do the conversion, you begin with the following expression:

8 feet 9 inches

The whole unit of feet isn't useful; you need to convert the units as described in Chapter 6. There are 12 inches in a foot. You can convert the whole number (8 feet) to inches with this simple expression:

$$8 \text{ feet} \times \frac{12 \text{ inches}}{1 \text{ foot}} = 96 \text{ inches}$$

The result is 96 inches. Add that to 9 inches from the problem.

96 inches + 9 inches = 105 inches

The answer is 105 inches.

Another mixed fraction example

In a medical office, a person's height is often measured and recorded in inches (for example, 69 inches). Express this height as a mixed number (containing feet and inches).

An inch is 1/12 of a foot, so 69 inches is 69/12 of a foot.

$$69 \text{ inches} = \frac{69}{12} \text{ feet}$$

Break the fraction into two fractions. Notice that $^{60}\!/_{12}$ of a foot easily converts to 5 feet. The 9/12 of a foot is 9 inches.

$$\frac{69}{12} = \frac{60}{12} + \frac{9}{12} = 5 \text{ feet } 9 \text{ inches}$$

Yet another mixed fraction example

You work in industrial or assisted living food service. You have $15\frac{1}{2}$ cups of pecans. You need $\frac{1}{2}$ cup of pecans to make one dozen muffins. How many dozens can you make?

Start with the mixed fraction 15½.

$$15\frac{1}{2}$$

Start by turning the whole number into a fraction. How many 1/2 cups are in 15 cups? Multiply by 1, but in the form of an equivalent fraction (2/2).

$$15 \times \frac{2}{2} = \frac{30}{2}$$

Fifteen cups has 30 half-cups in it. Now add the converted whole number to the fraction (1/2).

$$\frac{30}{2} + \frac{1}{2} = \frac{31}{2}$$

You have 31 half-cups. Because you need one half-cup to make a dozen muffins, you can bake 31 dozen muffins.

Ratios

Fractions are related to ratios. A ratio is a way to compare two quantities relative to each other. A ratio is expressed as two numbers separated by a colon (:). For example, 3:4.

In a vinaigrette dressing, the classic ratio of oil to vinegar is three parts oil to one part vinegar, written as 3:1. You typically describe this as "a 3-to-1 mixture." This ratio of ingredients is true whether you're mixing ounces of ingredients at home or gallons in a large cafeteria.

A ratio is not as straightforward as a fraction. It describes the parts, but not the whole, so don't automatically trust ratios without understanding what they really represent.

There's nothing vulgar about decimals

The fractions you see in this chapter are called *vulgar fractions*. No, that doesn't mean they have bad manners; you can still take them out in public.

Vulgar is from Latin, and just means common, ordinary, or everyday. That applies to the fractions that have two numbers separated by a line.

The opposite of a vulgar fraction is a decimal. Decimals can be called *decimal fractions*. After all, a number with decimal places shows a whole part and a partial part (for example 3.54). The whole number is, of course, a whole number, but the decimal part is a portion (a fraction) of a whole number. The number 3.54 is another form of a mixed fraction.

For example, in small two-cycle engines, you want to mix fuel with oil for lubrication. Common mix ratios are 12:1, 16:1, 24:1, and 32:1. The ratio 12:1 tells you to mix 12 parts of gasoline with 1 part of oil. Here, the result is a total of 13 parts: 12 of gas and 1 of oil. So the final mix is 12/13 oil and 1/13 gas. Notice that these fractions are similar, but not identical, to the ratio. By using the ratio alone, you end up with the mixture you need, but knowing the exact fraction can also be valuable.

The same is true in professional mixology (bartending), where cocktails are expressed as ratios as well as whole and fractional quantities. The classic dry martini uses 2½ ounces gin and 1/2 ounce dry vermouth. This recipe is a ratio of 2½:1/2. Because fractions in ratios are a little clumsy, double both sides of the ratio and you get a 5:1 ratio of gin to vermouth.

You can turn ratios into fractions. Consider the cyberdating success ratio developed in 2005 by the University of Bath. For every 100 cyberdates (dates with people who met online), 94 went on to date each other again and 6 did not. This figure is a 94:6 ratio. You can express the proportion of people dating again as 94/100 and those not dating again as 6/100.

Performing Math Operations with Fractions

Adding, subtracting, multiplying, and dividing regular numbers isn't hard. Why should fractions be any harder? They aren't. Any difficulties are usually with the bottom number, and that's easy to fix.

Mathematicians use some complicated math processes for fractions (for example, the rationalization of monomial and binomial square roots in the denominators of fractions), but you aren't likely to need them in your work.

Most jobs call for the simplest forms of fraction math. (***Note:*** Though addition and subtraction are typically the most basic math operations, we start with multiplication and division here because you actually need those skills to add and subtract.)

Multiplying fractions

Multiplying fractions is the easiest of the fraction math operations. Just multiply the numerators and then multiply the denominators.

To solve this problem:

$$\frac{3}{8} \times \frac{2}{3}$$

follow these steps:

1. **Multiply the numerators and the denominators.**

$$\frac{3 \times 2}{8 \times 3} = \frac{6}{24}$$

2. **Reduce the result, if you can.**

You can divide the top and bottom of 6/24 by 6. This gives you 1/4.

Multiplying a fraction by an equivalent fraction

You can multiply the numerator and denominator of a fraction by the same number, as long as it's not 0. A zero in the denominator produces bad, nonsensical results. You can multiply by an *equivalent fraction* — that is, a fraction with the same number in the numerator and the denominator — because it's really multiplying by 1. As shown in the following example, when multiplying by 3/3 (which is equal to 1), you simply get a "larger" version of the same fraction.

$$\frac{3}{4} \times \frac{3}{3} = \frac{9}{12}$$

The new fraction has the same value as the original fraction. Use this operation whenever you want to make denominators alike.

Multiplying a fraction by 1

In any kind of math problem, when you multiply anything by 1, the result is the original quantity. This holds true for fractions as well as for "regular" numbers.

$$\frac{3}{7} \times 1 = \frac{3}{7} \times \frac{1}{1} = \frac{3 \times 1}{7 \times 1} = \frac{3}{7}$$

Multiplying a fraction by 0

When you multiply anything by 0, the result is 0, and that's true for fractions. Notice in the following example that 3/7 is multiplied by 0/1, which is another way of writing 0.

$$\frac{3}{7} \times \frac{0}{1} = \frac{3 \times 0}{7 \times 1} = \frac{0}{7} = 0$$

This is the same as writing:

$$\frac{3}{7} \times 0 = 0$$

Dividing fractions

The hardest operation in fraction math is division. But life is short, so we make this operation easy.

In division, what you're dividing into is the *dividend,* the number doing the dividing is the *divisor,* and the result is the *quotient.* With whole numbers, 8 ÷ 4 = 2. The dividend is 8, 4 is the divisor, and 2 is the quotient.

The rule for dividing fractions is to invert the divisor and multiply. *Inverting* just means turning the divisor fraction (the second fraction) upside down. The bottom becomes the top and the top becomes the bottom.

Here's an example:

$$\frac{4}{7} \div \frac{2}{3}$$

1. **Invert the divisor.**

   2/3 becomes 3/2.

   $$\frac{4}{7} \times \frac{3}{2}$$

2. **Multiply the numerators and the denominators.**

   When you multiply 4/7 by 3/2, you get 12/14.

   $$\frac{4}{7} \times \frac{3}{2} = \frac{4 \times 3}{7 \times 2} = \frac{12}{14}$$

3. **Reduce the fraction, if you want, by dividing both the top and bottom by a common factor.**

   In the example, the terms can both be evenly divided by 2, which gives you 6/7.

Don't be taken in by story problems where they ask you to "divide by half." That's not the same as dividing by 2. For example, 10 divided by half is 10 ÷ 1/2 and the answer is 20. 10 ÷ 2 is 5.

Reducing a fraction

You can divide the top and bottom numbers of a fraction by the same number (an equivalent fraction). This is called *reducing* a fraction. Reducing a fraction usually makes the fraction easier for people to read and use.

If you multiply 3/7 by 7/6, you get a result of 21/42.

$$\frac{3}{7} \times \frac{7}{6} = \frac{3 \times 7}{7 \times 6} = \frac{21}{42}$$

But that number's too clumsy. So then you divide both the numerator and the denominator by the same number (in this case, by 21). You get a very nice, pretty, and reasonable 1/2.

$$\frac{21}{42} = \frac{1}{2}$$

Fractions as a representation of division

Be sure to remember that, in addition to dividing with fractions, fractions themselves can be used to indicate division. For example, when you see a fraction like 3/4, you can take it to mean "3 divided by 4."

The concept of fractions as division comes up more in Chapter 9, where we discuss decimals and converting fractions to decimals.

Adding fractions

The key to adding fractions is that their denominators (the bottom numbers) must be the same. When this is so, adding fractions is usually just like adding whole other numbers.

Adding fractions with the same denominator

The following fractions have the same denominator, 9:

$$\frac{1}{9} + \frac{5}{9} + \frac{7}{9}$$

To add the fractions, just add the numerators, as you can see in the following equations. Whether they are three fractions, one fraction with 1+5+7 combined, or the sum of 13/9, it's all the same. You have 13/9.

$$\frac{1}{9} + \frac{5}{9} + \frac{7}{9} = \frac{1+5+7}{9} = \frac{13}{9}$$

Converting to a mixed number (1⅘) is optional:

$$\frac{13}{9} = 1\frac{4}{9}$$

Adding fractions with different denominators

When denominators are the different, you must make them the same. One of the great misquotes from the movies is "We have ways of making you talk." Well, apply it here: "We have ways of making denominators the same" (bwa-ha-ha).

To make the denominators the same, find a *common denominator,* a number that both denominators can be converted to.

In this example, you want to add 3/4 and 2/3.

$$\frac{3}{4} + \frac{2}{3}$$

This addition is easy if you convert each fraction to have the same denominator. You need to find a denominator that you can use to convert 3/4 and 2/3. To find the common denominator, just multiply the two bottom numbers, 4 and 3. The answer is 12, and you can use that as the denominator. (Any other number that's a multiple of both 4 and 3 works, but 12 is the easiest to identify.)

Follow these steps:

1. **Multiply the top and bottom of the first fraction by the equivalent fraction that results in the common denominator.**

   In this example you multiply the top and bottom of 3/4 by 3 (which is okay because you're actually multiplying the whole fraction by 3/3, which is equal to 1). You get the answer 9/12.

$$\frac{3}{4} \times \frac{3}{3} = \frac{9}{12}$$

2. **Repeat Step 1 for the second fraction, using a different equivalent fraction that results in the same common denominator.**

   Multiply the top and bottom of 2/3 by 4, giving 8/12.

   $$\frac{2}{3} \times \frac{4}{4} = \frac{8}{12}$$

3. **Using the new fractions formed in Steps 1 and 2, add the numerators.**

   Add 9 and 8 together, giving 17/12.

   $$\frac{9}{12} + \frac{8}{12} = \frac{9+8}{12} = \frac{17}{12}$$

4. **If you choose, convert to a mixed number when appropriate.**

   17/12 can be converted to $1\frac{5}{12}$ if you want.

Add fractions with the same denominator. If the denominators aren't the same, make them the same.

Subtracting fractions

As with addition (see "Adding fractions" earlier in the chapter), the key to subtracting fractions is that their denominators (the bottom numbers) must be the same. The following sections show you how.

Subtracting fractions with the same denominator

When the terms in a subtraction problem have the same denominator, as you see in the following equation, you use pretty straightforward subtraction in the numerator:

$$\frac{7}{9} - \frac{5}{9} - \frac{1}{9} = \frac{7-5-1}{9} = \frac{1}{9}$$

Just subtract the numerators 7 – 5 – 1 to get 1. The answer is 1/9.

Subtracting fractions with different denominators

Like adding fractions, when you're subtracting fractions, you must make the denominators the same.

The following example is similar to the earlier addition example.

$$\frac{3}{4} - \frac{2}{3}$$

Follow these steps:

1. **Multiply the top and bottom of the first fraction by the equivalent fraction that results in the common denominator.**

   Multiply the top and bottom of 3/4 by 3 (which is okay because you're actually multiplying the fraction by 3/3, which is equal to 1), giving 9/12.

   $$\frac{3}{4} \times \frac{3}{3} = \frac{9}{12}$$

2. **Do the same thing to the second fraction, using a different equivalent fraction that results in the same common denominator.**

   Multiply the top and bottom of 2/3 by 4/4, giving 8/12.

   $$\frac{2}{3} \times \frac{4}{4} = \frac{8}{12}$$

3. **Using the new fractions formed in Steps 1 and 2, subtract the numerators.**

   Subtract 8/12 from 9/12, giving 1/12.

   $$\frac{9}{12} - \frac{8}{12} = \frac{9-8}{12} = \frac{1}{12}$$

Subtract fractions with the same denominator. If the denominators aren't the same, make them the same.

Example: Dividing and Selling a Cheesecake

You have a 3-pound (48-ounce) cheesecake, to be sold as an upscale dessert. You want to divide it into 12 slices. What fraction of the whole cheesecake is each piece? How many ounces does each piece weigh?

The first question is easy to answer, if you examine it. This is a simplified example of a math concept called *inspection*. If you cut the cheesecake into 12 slices, each slice is 1/12 of the cheesecake.

Here is a representation of the problem as a fraction.

$$\frac{1 \text{ cheesecake}}{12 \text{ slices}} = \frac{1}{12} \text{cheesecake/slice}$$

But how many ounces are in each slice? Think in ounces, not pounds. There are 48 ounces in 12 slices. If you express this amount as a fraction and simplify the numerator and denominator (that is, divide them both by 4), you see

$$\frac{48 \text{ ounces}}{12 \text{ slices}} = \frac{4 \text{ ounces}}{1 \text{ slice}}$$

When you divide 48 ounces of cheesecake by 12 slices, you get ounces per slice, which in this case is 4 ounces.

Pricing your cake wholesale

The cheesecake costs $24 wholesale. Given that one slice is 1/12 of the cake (as determined earlier), what is the wholesale cost of each slice?

Just as you can divide slices into weight, you can divide slices into dollars.

$$\frac{24 \text{ dollars}}{12 \text{ slices}} = 2 \text{ dollars/slice}$$

The cheesecake has a wholesale price of $2 per slice.

Pricing your cake retail

You want to sell slices of cheesecake for two times (2/1 times) the wholesale price (see the preceding section). What should the retail price be?

Express the wholesale price as a fraction and then multiply by a fraction representing what you want to sell the slices for, using 1 as the invisible denominator.

$$\frac{2 \text{ dollars}}{1 \text{ slice}} \times \frac{2}{1} = 4 \text{ dollars/slice}$$

When you take a $2 wholesale slice and sell it for two times as much, it's a $4 retail slice. You can just as easily retail the cheesecake for three times (3/1 times) or four times (4/1 times) the wholesale price, if it's an amaretto cheesecake with a toasted almond crust, finished with whipped cream and a squirt of chocolate sauce.

This concept is a small part of "budgeted cost percentage" evaluations that new chefs learn. The process is more elaborate when costing all ingredients and other factors that go into preparation of a dish.

Example: Cutting Fire Stops for Framing Carpentry

You are framing a room with 2 x 4 studs on 16-inch centers. How long should the fire stops be? (A *fire stop* is a short piece of 2 x 4 nailed between studs. Its job is to retard the spread of fire).

This problem uses a great story-problem-solving technique: drawing a picture. In Figure 8-7, you can see what studs on 16-inch centers look like. Carpentry experience (or a trip to the Internet) tells you that a 2-x-4 stud isn't 2 inches wide on the edge; it's 1½ inches.

So the length of the fire stop must be 16 inches, minus half the thickness of one stud and minus half the thickness of the other stud. Half the thickness of 1½ inches is 3/4 inch, so write the following equation that describes everything you need to do:

$$16 - \frac{3}{4} - \frac{3}{4} = 14\frac{1}{2}$$

That is, take the 16-inch distance and subtract 3/4 inch. Then take what's left and subtract another 3/4 inch. The answer is 14½ inches.

Bonus example: Suppose you're using pre-cut 8-foot wall studs (which are really only 92⅝ inches long) to make the fire stops. How many fire stops can you cut from one stud?

One of the best techniques for addressing story problems is to look for common sense possibilities. You can easily see that one fire stop is too few stops because it only takes 14½ inches off the stud you're cutting from. And ten fire stops is too many because that's over 140 inches, and the stud is only 92 inches and change long.

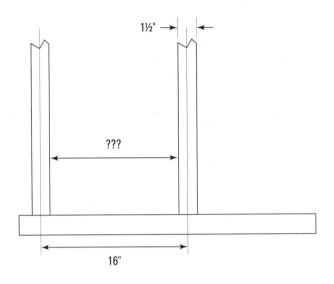

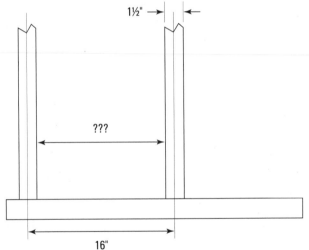

Figure 8-7:
Calculating
the length of
a fire stop.

Try the technique of *guessing*. You don't need a precise answer; you just need to come close. If there's any wood left over from this stud, just toss it on the waste pile. Follow these steps to use guessing for this example:

1. Multiply the length of one fire stop (14½ inches) by five.

You can start with whatever number you want, but because you know the answer here is somewhere between one and ten, five helps you eliminate half the numbers as being too high or too low right off the bat (assuming the answer isn't five itself).

$$5 \times 14\frac{1}{2} = 72\frac{1}{2}$$

Five stops take 72½ inches, but you still have plenty of stud left over. So this answer isn't big enough.

2. Make another guess a bit higher than five.

Maybe seven fire stops will fit.

$$7 \times 14\frac{1}{2} = 101\frac{1}{2}$$

Nope! Too many! This amount takes 101½ inches, which is more length than the stud has.

3. Try a number between your high and low guesses.

Between five and seven lies six. Okay! Six fire stops take 87 inches, with just a little waste. And that's the best answer you're going to get.

$$6 \times 14\frac{1}{2} = 87$$

If you want a fancy term for what you've just done, it's *successive approximation*, the technique of applying a series of values in a formula to develop a close approximation of a desired value.

Chapter 9

Decimals: They Have Their Place

In This Chapter

▶ Looking at the uses of the decimal point and decimal numbers

▶ Doing math with decimals

▶ Converting fractions to decimals and decimals to fractions

▶ Rounding decimals in your work

▶ Using decimals in change-making and figuring sales tax

Decimal numbers (sometimes just called *decimals*) are so commonplace that you probably hardly think about them. Like paved roads and running water, they show up every day in your work and your personal life. And like the roads and the water, you would have a hard time getting through a day without them.

You can't escape decimals in the digital age. If you're making doors, moldings, or signs or doing granite or stone work, you have to be an artist, a craftsperson, *and* a mathematician. The machine shop has used decimal fractions for decades. Also, the lab and the hospital use the metric system extensively, and all metric numbers are decimal numbers. Metric measurement is also increasingly the standard for working with cars and cooking.

Decimals and decimal fractions are easy. In this chapter, you review the names for decimal numbers and their parts, do math with decimals, and perform conversions. You also see handy tips about money and sales tax — all requiring decimal math.

Decimal numbers for modern times

Decimals are modern and precise, and that's what you need in modern times. In the ancient past, builders probably built just by judging the dimensions of stone and wood. Later, they measured, using sticks and knotted ropes. Fast forward to modern times. In classical carpentry, "vulgar" or common fractions (see Chapter 8) were the way to go. You built everything using feet, inches, and fractions like

$$\frac{1}{2}, \frac{3}{4}, \frac{7}{8} \text{ and } \frac{15}{16}$$

This scenario was true for both roughing in and for finish work, and the best guys (seems like they were all men then) could do the math in their heads.

But times change. There are no more 15¢ hamburgers, tacos, or gallons of gas. In fact, even the use of the cents sign (¢) has pretty much disappeared. So you can be sure this century's carpenter uses CAD/CAM software for design and machine programming, passes the data to a CNC router (or another woodworking tool), and watches the machine cut quickly and accurately. The software and the machines use decimal numbers.

Diving into Decimal Basics

The *decimal* number system is a base-10 system, with digits from 0 through 9. It's a *positional* system, where numbers appear in columns (for example, the 1s column, the 10s column, and the 100s column).

Zero (0) is a placeholder. You use it where a particular column contains nothing.

If you're an ancient Babylonian, you use a sexagesimal (base-60) system. You have also been dead for several thousand years, so you're looking pretty good. If you're of the Nunggubuyu people (in Numbulwar, Northern Territory, Australia), you use a base-5 system.

Whole numbers aren't much to write home about (see Chapter 3). However, when you work with decimal numbers smaller than 1, you're in the world of decimal fractions, and life gets interesting and more meaningful.

A *decimal fraction* is a decimal number whose value is more than 0 but less than one. The distinguishing marks of a decimal fraction are the decimal point and decimal places, which we cover in the following section. The following are examples of decimal fractions:

0.1   0.557   0.9999999999999999

A *mixed number* is a combination of a whole number and a decimal fraction. For example, the following are mixed numbers:

23.1   100.557   999,999,999.9999999999999999

The part of the number to the left of the decimal point is the whole number part, while the part of the number to the right of the decimal point is the fractional part.

Pointing out decimal points and places

The *decimal point* is a symbol that shows the boundary between the whole number part *(integer)* and the fractional number part *(decimal fraction)* of a number. You can also call the decimal point the *decimal separator*, but nobody outside of the Mathematics Department ever does. In the United States and many countries, the decimal separator is a period (.), but in many other countries, it's a comma (,). For example, in the United States, folks express a number as 12,345.67, but Germans use the comma, so the number is 12.345,67.

The following are numbers use the decimal point to show the separation between the whole part and the fractional part:

0.789   1.5   3.33 $4.95   −12.6

The first number, 0.789, doesn't have a whole number component. When that happens, the whole number part is understood to be 0. Use a 0 in front of the decimal point.

Decimal places are the digits to the right of the decimal point. They aren't stores that carry decimals on aisle 10. For example, the number *1.5* has one decimal place; the number *3.33* has two decimal places. Money ($4.95, for example) always has two decimal places, with the very minor exception of gasoline pricing and sales tax calculations.

The first number to the right of the decimal point is the *first decimal place.* The second number is the *second decimal place,* and so forth.

The value of each decimal place decreases as you move right. For example, 0.01 (two decimal places) is "one hundredth," or

$$\frac{1}{100}$$

while 0.001 (three decimal places) is "one thousandth," or

$$\frac{1}{1,000}$$

Many numbers have no decimal places (0 decimal places). They're *whole numbers*. For example, 23 is a whole number with no decimal places. You can write it with a decimal point (as 23.), but it's not necessary and nobody does it.

World War II soldiers were prone to a bit of decimal place humor. An old Army jokes says, "An M1 Garand rifle weighs 9.5 pounds. After you carry it a few miles, the decimal point falls out."

Precision, pennies, and parsing

Decimal points have three special uses:

- ✔ Indicating precision in calculations
- ✔ Expressing amounts of money
- ✔ Separating groups of numbers

Precisely, my dear Watson

Decimal places can express precision. Most people use *accurate* and *precise* as synonyms, but they don't mean the exactly same thing. The dictionary describes *accuracy* as "the degree of correctness of a quantity." *Precision* is "the degree to which the correctness of a quantity is expressed." Translation: You can measure something, get it wrong, but get identical results every time. Now, that's precise, even if it isn't accurate.

A machinist can mill a bar with a precision of 0.0001 inch (one ten-thousandth of an inch) again and again. That's precision. However, the whole job may be .25 inch off. Oops! That's an error in accuracy. The mill provides the precision, but the machinist sets the accuracy. For hunters and target shooters, imagine zeroing in a new rifle. If it shoots all the shots in a tight group next to each other, but they're all four inches low and four inches to the left, the rifle is precise but not accurate. Best to adjust those sights.

Be careful about numbers with trailing zeroes. For example, 2.3 and 2.300 are mathematically equal, but the extra zeroes imply that you measured the quantity to three decimal places. If you didn't, don't use the extra zeroes.

Life is filled with examples of precision. A laboratory scale easily weighs to three decimal places. A digital bathroom scale or a digital postal scale weighs to one decimal place.

Pennies are the root of all money

U.S. money is based on the dollar. Fractional dollars are based on hundredths. You call a hundredth of a dollar —

$$\frac{1}{100} \text{ dollar}$$

— a cent or a penny. In fact, a lot of countries denominate their money in units and hundredths of units.

Usually, you express dollars and cents with a decimal and two decimal places (for example $1.99) — *never* with one decimal place. For whole dollars, you use zero cents (for example, $1.00). But for whole dollars, you often leave off the cents and just write the dollar amount (for example, you write $23.00 as $23). You say this figure as "twenty-three dollars" or even "twenty-three bucks."

For deep and mysterious reasons known only to Big Oil, gasoline pricing is a little different. Every gas station in the known universe charges some amount plus

$$\frac{9}{10}$$

of a cent per gallon. For example, a price on a filling station sign is something like:

$$\$4.12\frac{9}{10}$$

The amount,

$$\frac{9}{10}$$

of a cent, is equal to 0.009 dollars, not quite a penny. The sign should read $4.129. No mathematician would accept a mixed number that contains integers, a decimal fraction, *and* a common fraction. However, everybody accepts this setup at the gas pump.

Parse the vegetables, please

Here's an example number that most people have difficulty reading: 123456789.55. The human mind wasn't designed to read 11 digits at once, even if they use a decimal point as a separator. Therefore, you usually parse larger numbers into smaller, more readable groups.

Often you see a number like the 11-digit example written as:

123,456,789.55

That's a lot easier to wrap your brain around. A comma (,) separates the units into groups of three. This system is what's used in the United States and many countries (including Australia, English Canada, Israel, Japan, and Malaysia).

But in countries that use the comma in place of the decimal point to separate whole and fractional numbers (as we discuss earlier in the chapter), the decimal point acts as the comma in parsing numbers. For example, in Germany, Belgium, Denmark, Italy, Romania, and much of Europe, you see:

123.456.789,55

You can find variations, too, which include spaces or apostrophes (') as separators.

In Microsoft Windows, you can alter the comma and decimal settings for your computer. Go to Start → Control Panel → Regional and Language Options. These options allow you change the region and use that region's conventions. You can also do some limited format customization for your region.

The Four Ops: Working with Decimals in Four Math Operations

In basic math, decimals are practically as easy to work with as integers are. You can do the math by hand, in a spreadsheet program, or with a pocket calculator or smartphone.

The discussion in this chapter is about The Four Ops (addition, subtraction, multiplication, and division — not to be confused with the Four Tops), but other operations — percentages, exponents and square roots — are just as easy.

If there's one rule for doing arithmetic with decimals, it's to *line up the decimal points* — it's vital for addition and subtraction. It's not as essential in multiplication and division, but just the same, staying neat and doing your housekeeping makes any math operation easier.

Adding excitement

To manually add decimals, try this method. If, for example, you want to add 12.695 and 3.02569, you can try it *inline,* in the form 12.695 + 3.02569. But that's the hard way. To make it easier, just stack the numbers:

12.695
3.02569

This arrangement is clumsy because nothing lines up. How do you know what to add to what? To fix the issue, line up the decimal points in the two numbers:

        12.695
        3.02569

Much better! If you're concerned that 12.695 doesn't have any numbers at the far right, just fill with zeroes:

        12.69500
        3.02569

Now add. The answer is 15.72069.

This method works for adding two numbers or many numbers. However, be careful on a calculator or smartphone. You may not have enough room in the calculator for a large number of decimal places.

With a spreadsheet, format the cells for several decimal places so that all the terms are sure to line up. The answer is the same regardless, but the lined-up version looks less confusing. Figure 9-1 shows both approaches on a spreadsheet, using the earlier example and two more numbers. The numbers in column C are much easier to read.

Figure 9-1:
Adding
decimals
by using
a spread-
sheet.

	A	B	C	D
1	1		1.00000	
2	7.1		7.10000	
3	12.695		12.69500	
4	3.02569		3.02569	
5	**15.72069**		**15.72069**	
6				
7				

Subtraction gives satisfaction

Subtraction is as easy as addition, and it should be. Subtraction is the opposite (or the *inverse*) of addition (see the preceding section). And the rules at the mathematics fitness center are the same for subtraction as for addition: Choose your method and line up the decimal points.

The addition example now becomes a subtraction example. Take 12.695 and subtract 3.02569 from it. You can write it inline as 12.695 – 3.02569, but that's no way to solve the problem. Instead, stack the numbers:

        12.695
        3.02569

Like addition, if you don't line up the decimal points, you run into trouble. So just line up the decimal points to save the hassle:

12.695
3.02569

Fill the top number with zeroes.

12.69500
3.02569

Subtract. The answer is 9.66931.

As with addition, note that you may not have enough room in a calculator or smartphone for a large number of decimal places.

Subtraction's easy with a spreadsheet. Like addition, format the cells so the decimal points line up. Figure 9-2 shows both approaches on a spreadsheet. The numbers in column C are much easier to read.

Figure 9-2:
Subtracting decimals by using a spreadsheet.

	A	B	C
1	12.695		12.69500
2	3.02569		3.02569
3	**9.66931**		**9.66931**
4			

Multiply with abandon

Multiplying decimals has as many wrinkles as an aging movie star. Fortunately, it has fewer side effects than plastic surgery.

You multiply decimals like you do "regular" numbers. Just don't lose track of the number of decimal places — it becomes important later in the process.

For example, if you multiply 3 by .25, the result is .75. The *multiplier* (number you're multiplying by) has two decimal places and the product also has two decimal places. The same is true if you have decimal places in the *multiplicand* (number being multiplied). For example, if you multiply 2.45 (with two decimal places) by 3, the result is 7.35, a product with two decimal places.

If both parts have decimal places, add up the total number of decimal places to determine how many places the product has. For example:

$1 \times 0.25 = 0.25$ 2 decimal places

$0.25 \times 0.25 = 0.0625$ 4 decimal places

$0.25 \times 0.25 \times 0.25 = 0.015625$ 6 decimal places

Here are some tips for multiplying decimals in each method:

✔ With a spreadsheet, just enter a formula: =0.25*0.25*0.25

✔ With a calculator or smartphone, just keep entering multipliers — $0.25 \times$ followed by $0.25 \times$ followed by 0.25 — and then press =.

✔ For manual work, lining up the zeroes doesn't get you anywhere. You multiply without regard to the decimal point(s) and then fix the number of decimal places at the end of the work. Just follow these steps:

1. **Stack the multipliers without worry about decimal alignment.**

   For example, to multiply 43.29 by 0.0265, simply write down the two numbers:

   43.29

   0.0265

2. **Multiply just as if the multipliers were integers.**

   In this example, your math looks something like this:

   $$\begin{array}{r} 43.29 \\ \times\,.0265 \\ \hline 21645 \\ 25974 \\ 8658 \\ 0000 \\ \hline 1147185 \end{array}$$

3. **Place the decimal point.**

   Because the multiplicand has two decimal places and the multiplier has four decimal places, set the decimal point six places from the right. The answer is 1.147185

But wait, there's more! Decimal multiplication gives you a little multiply-by-ten bonus. The rule is simple: To multiply a number by ten, just shift the decimal point one place to the right. To multiply the answer from the exercise (1.147185) by 10, shift the decimal:

$1.147185 \times 10 = 11.47185$

To multiply by 100, shift the decimal two places to the right, and so on.

Division is an important decision

Normal division of decimals isn't normal. It's the problem child of the Four Ops because the results have variations, but you'll come to love it.

Here are a few things to remember about dividing decimals:

- ✔ As with other operations, your calculator or smartphone may not have enough room to hold the answer. If you're not worried about the digits in the more miniscule decimal places, this limitation shouldn't cause a problem.

- ✔ With a spreadsheet, format the cells to allow for a large number of decimal places — about eight places is a safe number.

- ✔ With manual calculating, be prepared to shift the decimal point in the *divisor* (number you're dividing by) and the *dividend* (number you're dividing into), so the math looks more "normal." For example, you can express 3.00 ÷ .35 as 300 ÷ 35 by shifting the decimal point two places to the right in both the divisor and the dividend. The answer is 8.5714.

    When you are dividing a larger decimal into a smaller one, you can do the same thing. For example, you can express 1.04 ÷ .3.25 as 104 ÷ 325 by shifting the decimal point two places to the right in both the divisor and the dividend. The answer is 0.32.

You can expect three kinds of results when you divide decimals:

- ✔ When you divide a number by a bigger number, expect a smaller decimal. For example:

    1.75 ÷ 3 = .583333333

- ✔ When you divide a number by a smaller number, expect a bigger decimal. For example:

    6.8 ÷ 0.3 = 22.666666

- ✔ When you divide a number by certain other numbers, you may get an infinite series of repeating decimals. For example:

    1 ÷ 7 = 0.142857142857142857142857142857142857 . . .

    In this situation, be prepared to do some rounding (which we cover later in this chapter).

Like decimal multiplication, decimal division cuts you a break when you're dividing by ten. To divide a number by ten, just shift the decimal point one place to the left. Look at this example:

0.58 ÷ 10 = 0.058

To divide by 100, shift the decimal two places to the left, and so on.

Decimal Conversion

Common fractions are useful, but you often need to convert them to decimal numbers; machines with digital input, especially computers, are often fraction hostile.

Machinists and cabinetmakers using computer-aided design aren't the only folks affected by fraction intolerance. Graphic design work uses computers, and that means entering decimals. For example, if you're designing a cover for a book with a spine thickness of

$$\frac{15}{16}$$

inches, you need to make allowance for that dimension in your design. Convert

$$\frac{15}{16}$$

inches to 0.9375 inches for your art program to understand what you want.

Decimal numbers are very useful, to be sure, but they have their limits. Sometimes you need to convert them to fractions. For example, if you want to mail an item that weighs 0.5625 pounds, that's great, but the post office doesn't do decimal pounds. It does ounces. You need to convert 0.5625 pound to

$$\frac{9}{16}$$

pound, which is 9 ounces. The following sections clue you in on both conversion processes.

Converting fractions to decimals

Turning a fraction into a decimal number is easy. In fact, you probably know some conversions by heart. For example, the easiest conversions are

$$\frac{1}{4} = 0.25 \quad \frac{1}{2} = 0.5 \quad \frac{3}{4} = 0.75$$

The rule for converting fractions to decimals is to simply divide the denominator into the numerator to get the answer. In the first example, divide 1 by 4 to get 0.25. Use a calculator or smartphone, a spreadsheet program, or pencil and paper.

For example, to convert

$$\frac{15}{16}$$

to a decimal number, just divide 15 by 16.

$$15 \div 16 = 0.9375$$

The answer is 0.9375.

Your job or school probably celebrates Act Like A Sumerian Day. No? Too bad! Well, if you want observe it anyway, you can do your conversions on a clay tablet, just like the Sumerians did in the good old (really old) days. In the 4th century BC, Sumerian scribes wrote characters on wet clay tablets with a reed stylus. No trees were destroyed, and they could recycle the tablets by soaking them in water.

Early in their careers, machinists learn basic decimal conversions for selecting drill bits. Many tables show drill bit diameters as fractions with their decimal equivalents, ranging from

$$\frac{1}{64}$$

inches to

$$\frac{63}{64}$$

inches. However, the world is changing. Fractional bit sizes are still common in America, but most of the rest of the world now uses metric sizes.

Converting decimals to fractions

Converting a decimal number to a fraction is just as easy as going from a fraction to a decimal (see the preceding section). Maybe easier. Try it for yourself and see if you can't do it in 0.5 of the time. If that sounds stupid, that's because it should be "half the time." That number *0.5* needs a decimal-to-fraction conversion.

The conversion rule is simple: Set the decimal up as a fraction and reduce it to its simplest terms. For example, to convert 0.75 to a fraction, set up the fraction as follows:

$$0.75 = \frac{75}{100}$$

The basic answer is

$$\frac{75}{100}$$

To reduce the fraction, divide both the top and bottom by a common factor. Decimal fractions always have 10, 100, 1,000, 10,000, and so forth as a denominator, so try dividing by factors of 10, such as 5 or 2. You may be able to do more division to get to the simplest terms. The most reduced answer is:

$$\frac{3}{4}$$

You can find greatest common factor calculators on the Internet with a quick search.

How do you know how big to make the denominator? The number of decimal places in the decimal number you have to convert tells you how many zeroes to use.

1 decimal place	$= \frac{1}{10}$	(one zero)
2 decimal places	$= \frac{1}{100}$	(two zeroes)
3 decimal places	$= \frac{1}{1,000}$	(three zeroes)
4 decimal places	$= \frac{1}{10,000}$	(four zeroes)
5 decimal places	$= \frac{1}{100,000}$	(five zeroes)

Round, Round, Get Around, I Get Around

Sometimes, the answer to a decimal calculation isn't useful because it has too many decimal places. In this situation, you need to do some *rounding*, replacing the answer with another value that's very close to the original. You *round up* and *round down*, depending on the original answer and the number of decimal places you want to round to.

Rounding is a common practice. You do it with money all the time and don't even think about it. For example, if you pay $5.98 for an item and someone asks you what it cost, you probably say, "Oh, about six dollars."

Here are the rules for rounding:

1. **Figure out how many places you want to round to.**

   With money, you normally round up or down to whole cents. That's two decimal places.

2. **Look at the numbers to the right of your chosen rounding place.**

   In the case of money, go to the numbers at the right of the two decimal places. For example, in the amount $10.876265, the numbers 6265 are at the right of the two decimal places.

3. **Round up or down depending on the number to the immediate right of your chosen rounding place.**

   If that digit is 5 or greater, round up to the next number. In this example, the third digit is 6, so you round up to $10.88.

   If the digit you're rounding from is less than 5, drop the remaining digits to round down. For example, if the amount to be rounded is $10.873265, the rounded answer is $10.87

You can round to any number of decimal places. For example pi (π) rounded to 2 decimal places is 3.14. Pi rounded to 5 decimal places is 3.14159. Just look at the number to the right of the place you want to round to.

Making Change and Charging Sales Tax

No matter what your career, the realities of business are that you buy and sell. If you buy for cash, you receive change in return. If you sell for cash, you're expected to give change. Either way, knowing how to make change accurately is a good idea.

As Ben Franklin noted, nothing is certain but death and taxes. In fact, almost all purchases (except some Internet sales) require paying sales tax, and sales tax computation is a common use of decimal numbers. The following sections show you how to do both of these calculations.

Making change

If you've ever gone out to a fast-food lunch, you've been on the receiving end of making change. Sometimes it's done well, but sometimes you see the clerk struggle; counter help in real life is, sadly, not always as good as the perky counter help in TV life.

Making change is the technique of returning to the customer the difference in money between the amount of a purchase and the money tendered. *Change* often refers mainly to loose coins but can also include paper money.

The modern school of change-making uses a no-math technique:

1. The cash register tells the clerk the charges.

2. You give the clerk money and he or she enters the "amount tendered."

3. The cash register tells the clerk what change to give you.

But what do you do if you work in a small business without cash registers? What do you do at the bake sales held by your church, your service club, or your child's soccer league? You count out change the way your grandmother did when she worked at Woolworth's! Say your customer gives you a ten-dollar bill for a $9.56 purchase. Leave his bill in plain sight on top of the cash drawer and follow these steps to count out his change:

You count his change out and give it back to him, describing what you're giving him. Start with the smallest coins, and don't give him all pennies. Say, "Your purchase was $9.56, out of $10.00. That's $9.56,

plus 4 cents (4 pennies) makes $9.60,

plus 5 cents (a nickel) makes $9.65,

plus 10 cents (a dime) makes $9.75,

plus 25 cents (a quarter) makes $10.00."

We authors don't make this stuff up! You may not believe it, but there's math problem called the *Frobenius coin problem* (named after German mathematician Ferdinand Georg Frobenius) which is about *not* being able to make change. The problem asks, "What's the largest amount of money that can't be counted out using only coins of specified denominations?" The answer depends on the coins in your specific problem. A variation of the Frobenius coin problem uses "McNugget numbers." McDonald's Chicken McNuggets were originally sold in boxes of 6, 9, and 20 nuggets, so the problem asks, "What's the largest number of McNuggets you can't make up by buying whole boxes?" Before McDonald's introduced the four-nugget Happy Meal, the answer was 43 nuggets, but now it's only 11.

Charging sales tax

Calculating sales tax is simple. This simplicity is a good thing because sales taxes themselves aren't simple anymore. For example, California has no one sales tax. At this writing, the sales tax is 8.875 percent in Nevada City, 8.375 percent in neighboring Grass Valley, and 9.75 percent in Oakland. To calculate and charge sales tax for a customer, follow these steps.

1. **Convert the sales tax rate from a percent to a decimal.**

   Just move the decimal point two places to the left. For example, 8.875 percent becomes 0.08875

2. **Multiply the amount of the purchase by the sales tax rate.**

   This calculation is simply decimal multiplication, which we discuss in "Multiply with abandon" earlier in the chapter. For a $1.00 purchase, the tax would be

   $1.00 \times 0.08875 = 0.08875

3. **Round your answer up.**

   For more on rounding, check out the earlier section "Round, Round, Get Around, I Get Around." $0.08875 becomes $.09

4. **Add the result from Step 3 to the purchase amount to get the final amount of the purchase.**

   $1.00 + $.09 = $1.09.

The math calculation in the example is correct, but it may not always reflect reality. Using a sales tax chart is a good idea. According to the California State Board of Equalization's chart, the tax on a $1.00 purchase in Nevada City is $0.08. The tax doesn't go to $0.09 until the purchase is $1.07.

Example: A Journey to Office Supply Heaven

You ask yourself why you ever gave up cooking to open your own restaurant. Gave up cooking? True statement, because now that you're the owner of Glenda's Chateau de Swell Eats, you cook less than ever. You may plan the menu, but your days are spent running the business. Today is no exception, because you're going to the big box office supply store.

You need to buy a ream of paper, a dozen pens, and a binder. Your local sales tax is 8 percent, and you have a 10-percent discount coupon. Table 9-1 shows you how much your shopping list items cost, the discount, and the tax rate.

Table 9-1	Figures for Restaurant Shopping Trip
Item	*Cost/percentage*
One ream of color inkjet paper	$8.79
One dozen pens	$17.99
One 3-inch three-ring binder	$5.99
Coupon	10% off
Sales tax	8%

Calculate the cost of your total purchase, applying the discount and the sales tax. Here's a step-by-step approach:

1. **Add the items to be purchased.**

   Use decimal addition.

   ```
    $8.79
   $17.99
    $5.99
   $32.77
   ```

   The total is $32.77.

2. **Calculate the 10-percent discount.**

   You first have to convert 10 percent to a decimal (0.1), and then multiply it by your total:

   ```
   $32.77
   ×0.10
   $3.277
   ```

   Adjust the result for the total number of decimal places in the multipliers. That is, the answer should have three decimal places, so your answer is $3.277.

3. **Round the discount to two decimal places.**

   This task calls for decimal rounding. $3.277 becomes $3.28

4. **Subtract the discount from the purchase price.**

   Use decimal subtraction.

   ```
   $32.77
   −$3.28
   $29.49
   ```

5. **Calculate the sales tax on the discounted amount.**

   Express the tax of 8 percent as .08. This calculation requires decimal multiplication.

   $29.49
   ×0.08
   $2.3592

   Adjust the result for the total number of decimal places. That is, the answer should have four decimal places. The answer is $2.3592.

6. **Round the discount to two decimal places.**

   This action calls for decimal rounding. $2.3592 becomes $2.36.

7. **Add the tax to the purchase to get the total.**

   $29.49
   $2.36
   $31.85

   The answer is $31.85.

Chapter 10

Playing with Percentages

In This Chapter

▶ Defining a percentage

▶ Checking out four frequently used conversions

▶ Performing basic percentage math

The words *percent* and *percentage* are everywhere. No matter your occupation, you most likely encounter percentages, both as part of your day-to-day tasks and as part of the business side, including buying things, selling things, and paying people. Percentages are also part of common talk and always make appearances in the media, which is just fine when you understand what percentages are and how to use them. But if you don't, you can end up with some big errors and distorted information.

In this chapter, you find out what a percentage is and how to convert numbers to percentages. As a bonus, you convert percentages to numbers. And you even discover some shortcuts, too.

Pinpointing Percentages: Half a Glass Is Still 50 Percent Full

You take half, I'll take half. Isn't that the same as saying we'll split something 50/50 or saying 50 percent for you and 50 percent for me? That's three ways of saying the same thing.

To put it simply, a *percentage* is a way of saying how large or small a quantity is compared to another quantity. It doesn't get a bit more complicated than that. (You can also call a percentage a *percent* or a *per cent*.) For the simplest things in life, simple words like "half" work fine. The old saying asks, "Is the glass half full or half empty?" The optimist looks at how *much* is in the glass. The pessimist looks at how *little* is in it. And your common sense tells you it's the same proportion, no matter what.

But life isn't usually about the simplest things. If you must prepare a solution that contains 1.64 percent of 6 percent sodium hypochlorite (NaOCl) for periodontal patient home care, simple doesn't work anymore. Of course, when you know the math, percentages get a lot simpler.

Simple or complex, percentages get the job done. You typically use a percentage in three ways:

- **To compare a quantity to the whole:** A D5W (normal saline) IV is 0.9 percent sodium chloride — 9 grams of sodium chloride (NaCl) dissolved in 1 liter of water.

- **To compare one quantity to another:** The solution 2/3D & 1/3S (3.3 percent dextrose / 0.3 percent saline) shows in its name the relative amounts of dextrorotatory glucose, sodium ions, and chloride ions in the solution.

- **To compare a quantity to an increased or decreased amount of the same quantity:** For example, a caterer may make allowance for 5 percent extra servings for a big dinner. A discount is a good example of a decreased amount (for example, 10 percent off a $1 item means it's selling for $0.90.)

One big advantage of a percentage is that it applies to any quantity. For example, an intravenous solution of 5 percent dextrose is 5 percent dextrose, whether it's in a 1,000-milliliter bag in a hospital or being manufactured in a 10,000-liter tank.

Some quantities can't exceed 100 percent. For example, if your gas tank were 110 percent full, you'd be spilling the extra 10 percent on the ground (and with today's gas prices, you may not want to do that). And regardless of what your coach or project leader says, you can't give 110 percent effort. She'll have to settle for 100. But in other situations, you may legitimately encounter percentages higher than 100 percent. A class enrollment can be 200 percent of the previous year's enrollment. A recipe can be scaled to be 150 percent of its original quantities.

The following sections give a couple of important concepts to remember about percentages.

A percentage is a fraction, but the denominator never changes

A percentage is a fraction. What makes it special is that the dominator is always 100. (*Percent* or *per cent* means "per centum." That's from the Latin phrase meaning "by the hundred.")

So, for example, 17 percent is

$$\frac{17}{100},$$

a fraction with 100 in the denominator. You use the expression one half —

$$\frac{1}{2}$$

— but when you convert that to a fraction with a denominator of 100, it becomes

$$\frac{50}{100},$$

or 50 percent.

What about larger percentages? You can write 300 percent as:

$$\frac{300}{100}$$

That doesn't look very pretty, but you can reduce it to 3. Yes, 3 is still a fraction —

$$\frac{3}{1}$$

— but it's not really eligible to be a percentage until the denominator returns to 100.

You can also have fractional percentages. If you have 100 items and then take $\frac{1}{2}$ of one of those items, you have $\frac{1}{2}$ of a 100th (because one whole item out of the hundred would be one one-hundreth), or

$$\frac{0.5}{100}$$

That's

$$\frac{1}{2}$$

percent or 0.5 percent.

A percentage is a ratio, too

A percent is a fraction. A fraction is a ratio. Therefore, a percent is also a ratio. Check out Chapter 8 for more information about ratios.

Here's a simple example. If you stock 100 cans of motor oil, and 40 of them are SAE 5W-20, then 40 out of 100 cans are 5W-20. That's

$$\frac{40}{100}$$

of the cans, or 40 percent of them. You have a 40:60 ratio of 5W-20 to other kinds of motor oil in stock.

When you have a few friends over for a hot evening of talking math, you can refer to a percentage as a *dimensionless proportionality*. A quantity is dimensionless when it doesn't have a physical unit. The parts of a percent form a proportionality because they have a constant ratio.

Percentages Are Good Converts

When the going gets tough, the tough convert from one form to another. You convert percentages into other numbers and other numbers into percentages when doing so is convenient. Or when you're just in the mood.

We cover these conversions in the following sections:

- ✔ Percentage to decimal
- ✔ Decimal to percent
- ✔ Percentage to fraction
- ✔ Fraction to percent

If the conversions in the following sections all seem circular, that's because they are. Fractions to decimals to percents and back again. Percents to fractions to decimals and back again. Decimals to fractions to percents and back again. They're all interrelated.

Converting percentages to decimals

You can do some percentage conversions in your head, but when you need to get serious, convert then to decimal numbers by using a calculator, a spreadsheet, or good old pencil and paper.

The rule for going from percentages to decimals is simple: Divide the percentage by 100. The result is a decimal number because the denominator in a percent is always 100 (as we discuss in "A percentage is a fraction, but the denominator never changes" earlier in the chapter). So (for example) 67 percent becomes

$$\frac{67}{100}$$

or 0.67.

If you need 20 percent of a 1 liter solution, you need

$$\frac{20}{100}$$

of it or 0.2 liters. You don't even have to plug the numbers into a calculator. To divide by 100, you can just move the decimal point two places to the left of its original position. The number 20 really is 20.0. When you move the decimal to the left, the result is 0.200 liters, or 0.2 liters — the same answer you got with the actual division.

What do you do when you don't see a decimal point? Just remember that the decimal point is always assumed to follow the ones position in the number. And, if the percentage itself has a decimal point, such as 37.6 percent, just shift the decimal two places to the left. The answer is .376.

Turning decimals into percentages

Converting from a decimal number to a percentage is just the opposite of converting from a percentage to a decimal (see the preceding section). When it comes to decimal conversions, you're faster than any calculator or spreadsheet.

To make this conversion, multiply the decimal by 100 and add a percent sign. The result is a percentage. For instance, if you want to make 0.67 a percentage, multiply 0.67 by 100 to get 67 and then slap a percent sign on it. As a shortcut, you can simply shift the decimal point two places to the right, which is the same as multiplying by 100.

If you machine 0.42689 inches of stock off a 1-inch aluminum bar, what percentage do you remove? To find out, just multiply by 100: $0.42689 \times 100 =$ 42.689 percent. You took off 42.689 percent of it.

Going from percentages to fractions

This conversion has a great taste and it's less filling. The rule is to drop the percent sign and put the number in a fraction over 100. The result is (naturally) a fraction. Reduce as necessary. This conversion comes in handy when working with a fraction is easier than working with a decimal. (Head to the earlier section "Converting percentages to decimals" for details on that calculation.)

For, example to convert 26 percent to a fraction, just follow these steps:

1. **Drop the percent sign and use the numerical portion of the term.**

   The 26 percent is just 26.

2. **Set the percentage number as a numerator over a denominator.**

   The denominator is always 100

   $$\frac{26}{100}$$

3. **Simplify your fraction.**

   $$\frac{26}{100} = \frac{13}{50}$$

The answer is

$$\frac{13}{50}$$

Transforming fractions to percentages

This conversion is the reverse of the conversion of percentages to fractions in the preceding section. The rule for converting a fraction to a percentage is to convert the fraction into a decimal and multiply by 100. The result is a percentage. For example, if you have

$$\frac{33}{100}$$

of something, you divide 33 by 100 to get 0.33 and then multiply by 100 to arrive at 33 percent.

What if you have ⅝ of something? First, divide 5 by 8 to get the decimal 0.625. Multiply by 100. The result is 62.5 percent.

Calculating Percentage Increases and Decreases

The rules for calculating percentage increases and decreases are very simple even without the available shortcuts. The following sections present the lightning round of increasing and decreasing percentages.

Percentage increases: You get 10 percent more!

To increase a number by a percentage, multiply the original number by the percentage and then add the result to that original base amount. For example, if you normally supply customers with a $50.00 item and must increase its price by 15 percent, multiply $50.00 by 15 percent ($50 \times 0.15$) to get $7.50. Add the result to the base amount to reach the new price: $50.00 + $7.50 = $57.50.

Of course, the same procedure holds true in calculating the tip in a restaurant.

Here's a tip about tips. To calculate a tip quickly, take the bill and figure 10 percent in your head. For a $50.00 dinner, that's $5.00. Then figure 50 percent of that figure (in this case, you get $2.50). Fifty percent of 10 percent is another 5 percent, and together they make 15 percent ($7.50). So $7.50 is a 15 percent tip for a $50.00 dinner. You also may find a tip calculator on your cellphone or smartphone.

Percentage decreases: You save 10 percent!

To decrease a number by a percentage, multiply that number by the percentage and then subtract the result from the base amount. For example, if you have a coupon for 20 percent off, and you want to buy a $50 item for your business, multiply $50 by 20 percent (0.20) to get $10. That's the amount of your discount. Subtract the result from the base amount ($50 – $10 = $40) to get your final price.

In the language of office supplies and fashion, "10 percent off" means the merchandise's regular selling price is discounted by 10 percent.

An advertising slogan such as "Discounts up to 20 percent or more" is totally bogus. The language doesn't mean anything. If the discounts are "up to 20 percent," your math lets you calculate any discount from 1 percent to 20 percent. But what about "or more?" If the discounts go above 20 percent, why didn't the merchant say so?

The 100 percent increase: You must be 100 percent satisfied!

To calculate a 100 percent increase in something, simply double the base amount. To calculate a 200 percent increase in something, simply triple the base amount.

"Wait, what?" you may be saying. "200 percent is twice something, not triple it!" That's correct, but in this case, you're taking a 200-percent *increase* — you have to account for the original amount as well. Because you're doubling the base and then adding it in again, a 200-percent increase therefore *triples* the amount.

Dividing a Pie Using Percentages

When you have a set of percentages, everything is easy. You divide something into parts by multiplying by a percentage.

"You divide by multiplying" has kind of a Zen ring to it, doesn't it? Why does it work? A percentage is a fraction, so multiplying a large quantity by a percentage has the effect of multiplying it by a fraction (which divides it into portions).

The logic of percentages works with both quantities and amounts. It can apply to marbles, pills, cans of motor oil, pounds of rice, and so on.

Your food service sells apple pies, and each slice is 12.5 percent of a whole pie. How do you divide the pie in the easiest way? Why, you just follow these easy pie-dividing steps:

1. **Turn the number into a fraction.**

   Follow the conversion rule to turn the percentage (12.5 percent) into

   $$\frac{12.5}{100}$$

2. **Reduce the first fraction by dividing both parts by a common factor (if possible).**

   In this example, you can divide the numerator and denominator by 12.5 to get

   $$\frac{1}{8}$$

   Cut the pie into eight equal slices.

Now here's an example of dividing by multiplying by two percentages. Your food service features two new portions of decadent pie, the Cholesterol Colossus (37.5 percent of a whole pie) and the Microscopic Minislice (6.25 percent of a whole pie). Your boss tells you that you can evenly slice up one pie with these portions. How do you divide the pie in the easiest way? Follow these steps:

1. **Turn the numbers into fractions.**

   Follow the conversion rule to turn the percentages into

   $$\frac{6.25}{100} \text{ and } \frac{37.5}{100}$$

2. **Reduce the first fraction by dividing both parts by a common factor (if possible).**

   In this example, you can divide the numerator and denominator by 6.25 to get

   $$\frac{1}{16}$$

3. **Repeat Step 2 to reduce the second fraction.**

   You can divide both parts by 12.5 to get

   $$\frac{3}{8}$$

4. **Figure how many giant slices you can get from the pie and how much (if any) of the pie is left over.**

   You can get two big slices (totaling

   $$\frac{6}{8}$$

   of the pie) but not 3 slices (because that would be

   $$\frac{9}{8}$$

which is more than the whole pie). This amount

$$\frac{6}{8}$$

is the same as

$$\frac{3}{4}$$

of the pie, leaving

$$\frac{1}{4}$$

of it.

5. **Determine how many of the skinny minislices you can get from the rest of the pie.**

   The remaining

   $$\frac{1}{4}$$

   pie is equal to

   $$\frac{4}{16}$$

   pie, so four of your

   $$\frac{1}{16}$$

   slices finish the pie off nicely.

6. **Divide the pie by cutting it into the number of slices you determine.**

   In this example, that's two

   $$\frac{3}{8}$$

   pie slices and four

   $$\frac{1}{16}$$

   pie slices.

About any sous chef or pastry chef knows about half, quarter, and eighth slices, so these measurements should be no problem. The ending fractions are good in the kitchen, while the starting percentages are good for the cost accountants.

Beware: Percentages can lie!

In a perfect world, the business information you use would be perfect and clear. But it's not, and that's another indication that it's not a perfect world. Various statistics can be distorted (whether accidentally or on purpose).

A New York Times columnist reported legitimate survey results with no hype: "Since 1996, the percentage of Americans who said that they have been in the presence of a ghost has doubled from 9 percent to 18 percent . . . ." A hard-working tabloid or cable news channel may distort these percentages with the grabbing headline "Number of people seeing ghosts up 100 percent!" A change from 9 percent to 18

percent is in fact a 100-percent change, but 9 or 18 percent is only a small percentage of people in a survey sample. It's dramatic, but not *that* dramatic.

When a sports team ends the season with 1 win, 1 tie, and 50 losses, the local news report may say "Bears won 50 percent of games they didn't lose!" That's true, because when you look only at wins and ties, the Bears had 1 win (50 percent) and 1 tie (50 percent). But it sure amounts to using percentages to distort the big picture (that they lost about 96 percent of their total games).

Example: The World of Pralines

You've lived in New Orleans all your life, and you have your grandmother's famous recipe for pralines. You sell about 90 dozen a day in a storefront in the French Quarter and online.

You visit your relatives in California. You want to make 3 dozen pralines, but you've long since forgotten the original recipe ingredient amounts; however, you do know your commercial amounts. Use percentages to scale the recipe to make the batch you need.

The recipe for 90 dozen (1,080) pralines is

- 30 cups granulated sugar
- 30 cups light brown sugar, packed
- 5.5 quarts half-and-half
- 7.5 teaspoons salt
- 2 pounds butter
- 5 ounces vanilla
- 7.5 pounds chopped pecans

These amounts are in units you'd use in a home kitchen. A commercial operation would use pounds of sugar, gallons of half-and-half, and tablespoons or ounces to measure salt. Although some of the units here may seem a bit unconventional, they make the math easier for the purposes of this example.

1. **Determine the percentage decrease from the recipe quantity (90 dozen) to the desired quantity (3 dozen).**

   Divide 3 by 90 to get 0.03333.

2. **Multiply the decimal by 100 to convert the decimal to a percentage.**

   You can move the decimal point two places to the right or do the math longhand (or on a calculator) as follows:

   $0.03333 \times 100 = 3.333$ percent

3. **Multiply each ingredient by 3.333 percent (or 0.0333) to develop the decreased amounts.**

   Table 10-1 shows you each ingredients amounts before and after the conversion.

Table 10-1	Decreasing Ingredient Amounts	
Ingredient	**Amount Before Decrease**	**Amount After Decrease**
Granulated sugar	30 c.	0.9999 c.
Light brown sugar, packed	30 c.	0.9999 c.
Half-and-half	5.5 qt.	0.1833 qt.
Salt	7.5 tsp.	0.2500 tsp.
Butter	2 lbs.	0.0667 lb.
Vanilla	5 oz.	0.1667 oz.
Chopped pecans	7.5 lbs	0.2500 lb.

4. **Convert the decreased amounts from decimals into units more suited for home cooking.**

   In some cases, you're just converting decimals to fractions, but some ingredients actually change units, so watch out. For safety's sake, use standard conversion units from the Internet or a cookbook. Table 10-2 shows you the decreased amounts from Table 10-1 converted to more familiar-looking amounts.

Table 10-2	Converting Decreased Amounts to More-Familiar Units	
Ingredient	*Decreased Amount*	*Familiar Conversion*
Granulated sugar	0.9999 c.	1 c.
Light brown sugar, packed	0.9999 c.	1 c.
Half-and-half	0.1833 qt.	3/4 c.
Salt	0.2500 tsp.	1/4 tsp.
Butter	0.0667 lb.	2 Tbsp.
Vanilla	0.1667 oz.	1 tsp.
Chopped pecans	0.2500 lb.	1 c.

The recipe is fully converted. You should get your batch of 3 dozen pralines with no problem.

Pralines get their name from Cesar du Plessis-Praslin (1598–1675), a duke and a Marshall of France. He offered almonds coated in cooked sugar to famous women, with "love" in mind. ("Want some candy, little girl?") When the French settled in Louisiana, they substituted pecans for almonds. And of course, Praslin didn't invent the praline. It was created by his chef, Clément Lassagne. (No, they didn't name lasagna after him.)

Example: Oily to Bed and Oily to Rise

Your shop has 240 cans of motor oil. 120 are SAE 5W-30. 96 are SAE 10W-30, and 24 are SAE 60. What percentage of your total supply of cans does each oil type make up?

1. **Organize your info.**

   Making a table like this one is a great way to do just that.

Oil Viscosity Grade	*Quantity*
SAE 5W-30	120 cans
SAE 10W-30	96 cans
SAE 30	24 cans
Total	240 cans

2. **Calculate the percentage of your supply each grade comprises.**

   Divide each grade's quantity by the total supply to determine how much of the total quantity that type represents. For example, to calculate the percentage for 10W-30, divide those 96 cans by 240:

   $$\frac{96}{240} = 0.40 = 40 \text{ percent}$$

3. **When you complete the three conversions, put them in a table to check your work.**

   Use a table like the one that follows; when the table is complete, add up the percentages. They should total 100.

Grade	*Percentage of Total Supply*
SAE 5W-30	50 percent
SAE 10W-30	40 percent
SAE 30	10 percent
Total	100 percent

Chapter 11

Tackling Exponents and Square Roots

*M*athematics has two mysteries that don't deserve that status: exponents and square roots. Knowing the concepts is important because then the mystery disappears. Life is challenging enough without having a long list of mystery items.

Exponentiation is a mathematics word that means "applying an exponent to a base" or "raising a number to a power." Using a few small numbers as exponents to generate large and useful numbers may seem mysterious, but it's not so hard. The *square root* of a number is another number that when multiplied by itself yields the original number. The mystery here is that there doesn't seem to be any obvious way to intuitively find square roots.

It's not as though you never use exponents or square roots in your work. Exponents and square roots are often just disguised or used limitedly. Practical use of exponents is the world of square and cubic measure, laboratory work (using metric units), and information technology. The practical world of square roots is very narrow and includes finding the radius of a circle or the edge length of square areas.

In this chapter, you see the way exponents work and review their most useful applications. You also discover the simple shortcuts in exponent math. The bonus part is that you see three ways to calculate a square root.

Exponentiation: The Power of Powers

Exponentiation is a mathematics process represented in expressions and equations by just a couple of symbols. After you've met up with the word exponentiation (and this chapter is your opportunity), you aren't likely to use it in everyday speech.

The basics of the base

An expression in exponentiation has two parts, a base and a power. The power is often called the *exponent*:

base^{power} or base^{exponent}

The *base* is a number you raise to a power. The *power* is a number you raise the base by (and it's indicated by a raised number — a *superscript*). Here's an example with specific numbers:

3^4

In the example, 3 is the base and 4 is the exponent. You say this number as "three raised to the fourth power" or "three to the fourth."

Some exponents have special names. You have seen numbers like these:

7^2 or 1,549 ft^2

When you raise a number to the second power (power of 2), you often call it *square measure*. Refer to the first example (7^2) as "seven squared." Say the second example (1,549 ft^2) as "1,549 square feet." "Square feet" is the common way of expressing the size of spaces or areas such as offices, parking lots, driveways, and carpeting.

When you raise a number to the third power (the power of 3) you often call it *cubic measure*.

4^3

You call 4^3 "four cubed." A *cubic foot* (CF) of volume is the volume of a container 1 foot in length, 1 foot in width, and 1 foot in height.

1 ft $\times$ 1 ft $\times$ 1 ft = 1 ft^3

"Cubic feet" is a common measure of commercial and consumer refrigerator capacity. When welders talk about 40 CF, 60 CF, and 80 CF tanks, they aren't referring to the size of the tank but rather to the amount of compressed welding gas the tank holds.

In the world of science, you may see metric volume measurements such as cubic centimeters (cm^3); however, in the lab or hospital, a cubic centimeter may be abbreviated as *cc* or referred as a *milliliter* (ml).

Moving beyond 2 or 3

Powers don't stop with just the second and third power. The most ambitious powers want to be bigger than 2 or 3 because higher powers (in addition to being generally revered) are a very compact way of expressing very large and very small numbers, which we discuss further in the "Powers with base 10" and "Powers with base 2" sections later in this chapter.

Table 11-1 shows you an example of a simple expression representing a large number:

Table 11-1	Conveying Large Numbers with Exponents	
Exponent	*Math*	*Result*
2^2	2×2	4
2^3	$2 \times 2 \times 2$	8
2^4	$2 \times 2 \times 2 \times 2$	16
2^5	$2 \times 2 \times 2 \times 2 \times 2$	32
2^6	$2 \times 2 \times 2 \times 2 \times 2 \times 2$	64
2^7	$2 \times 2 \times 2 \times 2 \times 2 \times 2 \times 2$	128
2^{28}	$2 \times 2 \times 2 \times 2 \times 2 \times 2 \times 2 \times 2 \times 2 \times 2 \times 2 \times 2 \times 2 \times 2 \times 2 \times$ $2 \times 2 \times 2 \times 2 \times 2 \times 2 \times 2 \times 2 \times 2 \times 2 \times 2 \times 2 \times 2$	268,435,456

In Table 11-1, the base (2) doesn't change, but the exponent does. Notice how quickly a small three-character expression (2^{28}) represents a very large number. Such large numbers are easier to manipulate when you express them in such a compact form. The same is true with SI units (the International System of Units, which you commonly call the metric system). A little bit of writing gets you a whole lot of number. The range of metric terms (for liters, for example) goes from the very tiny *yoctoliter* (10^{-24}, or 1/1,000,000,000,000,000,000,000,000) to the very large *yottaliter* (10^{24}, or 1,000,000,000,000,000,000,000,000).

Different faces of special bases

When you work with exponentiation, you can use almost any base and almost any power. In algebra, you see terms whose values aren't known — for example, a^b — and they can represent anything until you know the solution.

However, you need to know about a few special bases and powers. A couple of them drive the world these days, and others are just oddities (unless you're a theoretical physicist). Here are tonight's contestants on *Dancing with the Power and Base Stars;* we discuss them in the following sections:

- ✔ **Powers with base 10:** Much like basketball and ice hockey power forwards, base 10 is the star of the show.

- ✔ **Powers with base 2:** This base has been rising fast since the dawn of the computer age.

- ✔ **Powers with base 1:** This dull base only has one trick.

- ✔ **Powers with base 0:** This base is also dull, but with a hint of controversy.

- ✔ **Powers with base –1:** A base with only two results.

- ✔ **Powers of 1 and 0:** These two don't deal with bases but with rather unusual properties of certain powers.

Powers with base 10

You grow up counting to 10. Later, you count to 100, and one day on a dull afternoon, you count to 1,000. These three numbers are all multiples of (and therefore powers of) 10. When you express them with powers, 10 becomes 10^1, 100 becomes 10^2, and 1,000 becomes 10^3.

In middle school and high school, you go a step further. Put a 1 in front of a 10 raised to a power and you have *scientific notation*. The numbers 10, 100, and 1,000 become 1×10^1, 1×10^2, and 1×10^3.

Powers of 10 are prominent in the lab and in the observatory, whether you measure the number of molecules in a reagent, the diameter of an atom's nucleus or the distance to a star. For example, NASA says the distance to Proxima Centauri is about 39,900,000,000,000 kilometers. It's a lot easier to write this distance as 39.9×10^{12} kilometers.

Using powers of 10 is a much easier way to do math on very large or very small numbers. In exponentiation, the equivalent of a decimal shift is increasing or decreasing the power. A one-decimal shift to the right gets you a tenfold increase in the value of a number. That is, to multiply a number with a base of 10 by 10, just increase its exponent by 1. A one-decimal shift to the left produces a tenfold decrease in the value of a number. To divide a number with a base of 10 by 10, decrease its exponent by 1. These shifts mean you're changing the *order of magnitude*.

Look at this example:

$$10 \times 1{,}000 = 10{,}000 \text{ or } 10 \times 10^3 = 10^4$$

That's increasing the number by a factor of 10.

You can also use negative powers of 10 — they represent division. A milliliter is $\frac{1}{1{,}000}$ of a liter, and you express that as 10^{-3} liters.

Powers with base 2

Powers with base 2 are at the very core of computing. From the earliest electronic computers to the latest ones you need for your work, the internal math is base 2 math.

Counting to ten was a pretty easy task for our ancestors, and civilizations have risen and fallen while they used decimal math. Trouble is, counting to ten is very complicated for a computer. Counting to 2, however, isn't bad. A microprocessor can multiply by 2 or divide by 2 just by doing a bit shift. Very efficient. If you're a microprocessor, you know two states (1 for "on" and 0 for "off"), and that's all you need.

You as a professional use the powers of base 2 in two ways:

✔ If you work with computers professionally (in information technology or as a computer tech), various base 2 numbers are your daily companions. Those numbers include

- Processor speed in Hz

- RAM in gigabytes

- Disk capacity in megabytes, gigabytes, and terabytes

- Network speed in megabits and gigabits

- Video board memory in megabytes

✔ If you buy or use computers as part of your work, you'll see base 2 numbers (or their abbreviations) all the time.

Table 11-2 lists some common abbreviations and their numeric representations in base 2 and base 10 (covered in the preceding section):

Table 11-2	Common Base 2 Computer Abbreviations	
Term	**Base 2**	**Base 10**
Megabyte	2^{20}	1,048,576
Gigabyte	2^{30}	1,073,741,824
Terabyte	2^{40}	1,099,511,627,776
IPv6 address space	2^{128}	340,282,366,920,938,463,463,374,607,431,768,211,456

Note that the name *megabyte* actually means 1 million bytes. However, the real number of bytes in a megabyte is 1,048,576, and everybody goes along with the convention.

Did you know that the world may be running out of Internet addresses? The current IPv4 system has only 4,294,967,296 (2^{32}) addresses available. The proposed IPv6 system would have 2^{128} addresses. That's a *big* number.

Powers with base 1

This topic is special. It won't come up in your work, but you need to know about it to get the whole picture. What happens when you elevate 1 to various powers? Use this handy table to see the answer:

Spoken Form	**Written Form**	**Math**	**Result**
1 to the first	1^1	1×1	1
1 squared	1^2	1×1	1
1 cubed	1^3	$1 \times 1 \times 1$	1
1 to the eighth power	1^8	$1 \times 1 \times 1 \times 1 \times 1 \times 1 \times 1 \times 1$	1

Yes, 1 to any power equals 1.

Powers with base 0

Powers with base 0 have produced a controversy, and you won't encounter them in your work. However, you need to know about them, so here are the rules, straight from multiple online sources.

✔ If the exponent is positive, the power of zero is zero:

$0^n = 0$, no matter how large the exponent.

✔ If the exponent is negative, the power of zero (0^n, where $n < 0$) is called "undefined," because division by zero is implied and that's impossible.

0^{-n} = undefined.

✔ If the exponent is zero, some mathematicians define it as one and others leave it undefined.

$0^0 = 1$ or may be undefined. Either choice is a safe bet for you.

Powers with base (–1)

A base of (–1) isn't much to shout about, but it's twice as exciting as powers with a base of 1 — it has two rules.

✔ When the exponent n even, $-1^n = 1$

✔ When the exponent n is odd, $-1^n = -1$

Powers of 1 and 0

Exponentiation has a couple of special conditions dealing with the powers (*not* the bases) of 1 and 0. This section introduces you to them, but the following section shows you them in action.

The math of exponentiation shows that any number raised to the power of 1 is itself, as the following examples show:

$1^1 = 1$

$10^1 = 10$

$756^1 = 756$

$2{,}568{,}145{,}259^1 = 2{,}568{,}145{,}259$

Any number raised to the power of 0 is 1, as indicated in the following examples:

$1^0 = 1$

$10^0 = 1$

$756^0 = 1$

$2{,}568{,}145{,}259^0 = 1$

Exponentiation math

Exponentiation math is fast, fair, and friendly. To multiply two numbers with the same base, you add the powers. To divide the numbers, *subtract* the powers. A term with an exponent represents repeated multiplication.

When you multiply two terms with exponents, you see that the result is identical to adding the exponents.

$$x^a \times x^b = x^{a+b}$$

For example, 5^2 is 5×5 and 5^3 is $5 \times 5 \times 5$.

That is, 5^2 is 25 and 5^3 is 125. If you multiply 5^2 by 5^3, you're multiplying 25 by 125. The answer is 3,125.

What if you add the exponents 2 and 3? The result is 5^5 or $5 \times 5 \times 5 \times 5 \times 5$. And the result of doing the multiplication is 3,125. The answer is the same: 3,125.

The same idea is true when you divide; you just subtract the powers.

$$\frac{x^a}{x^b} = x^{a-b}$$

Use the figures from the multiplication example and divide 5^3 by 5^2.

$$\frac{5^3}{5^2} = \frac{5 \times 5 \times 5}{5 \times 5} = \frac{125}{25} = 5$$

What if you just subtract the exponents? You get the same result.

$$\frac{5^3}{5^2} = 5^1 = 5$$

Now if the result turns out to be 5^0, the answer is 1, because any base raised to the power of 0 is 1.

If the result turns out to be a negative exponent (such as 5^{-1}), it's a *reciprocal*. The general rule is

$$x^{-a} = \frac{1}{x^a}$$

Using 5 as an example:

$$5^{-1} = \frac{1}{5^1} = \frac{1}{5}$$

The answer is

$$\frac{1}{5}$$

And to see this process in a longer form, divide 5^2 by 5^3.

$$\frac{5^2}{5^3} = \frac{5 \times 5}{5 \times 5 \times 5} = \frac{25}{125} = \frac{1}{5}$$

Again, the answer is

$$\frac{1}{5}$$

Getting Back to Your (Square) Roots

You know how to square numbers. The subject may have come up in elementary school very soon after you learned multiplication, and then again in high school math and maybe in community college math. And multiplying to get a square value comes up in several chapters in this book, including this very one. The earlier section "The basics of the base" shows you that a *square* is another way of describing a number raised to the power of 2. For example *5 squared* is also 5^2 and equals 25.

But what if you have the square and need to find out what it's the square of? The inverse of squaring a number is finding a number's *square root*. A square root is all about finding the value of the base when you know only the result of squaring the value. A square root operation has a symbol. The following symbol shows a problem where you're looking for the square root of 16:

$\sqrt{16}$ or $\sqrt{16}$

So what's the square root of 16? It's 4. When you were younger, you probably studied the easy ones: 4 (answer, 2), 9 (answer, 3), 16 (answer, 4) and 25 (answer, 5). That's great, but what happens when you need to find the square root of 625? You need a math solution, not memory. Luckily, the following sections give you three methods for finding square roots: the hard way, the easy way, and the effortless way.

Note: Truth in advertising: You may not find a lot of opportunities to use square root math in your trade. For example, if a cement mason knows the area of a circular patio, he or she can use a square root formula to find the patio's radius and diameter. But that's a bit backward because the mason usually starts with the linear dimensions (such as the radius) and then comes up with the area to calculate the volume of the concrete pour.

Why are they called roots, anyway? Maybe because, like a plant's roots, they're basic and lie below the surface. Two sources say that in math, the *root* of a number *x* is any number that, when repeatedly multiplied by itself, eventually yields *x*. A square root is a root where the number is multiplied by itself once. A cube root is where the number is multiplied by itself twice. Higher-order roots are possible, too.

Square roots the hard way

A manual method for finding square roots does exist, but it's not for the faint-hearted. High schools taught it in the 1960s, and even then only in the accelerated math classes. It's cumbersome and slow, although it's accurate. You can find the method on the Internet, but the chances are good that you won't find it to be a good use of your time.

Square roots the easy way

Use the technique of *successive approximation* (also called *guessing*) with the help of a calculator. This method is the thinking person's guessing.

For example, if you want to find the square root of 19 to three decimal places, take the following approach:

1. **Examine the situation.**

   You know that the square root of 25 is 5. The result 25 is too high, so the root 5 is too high. You know that the square root of 16 is 4. That result (16) is too low. The answer you want is somewhere between 4 and 5.

2. **Split the difference between your too-high and too-low numbers.**

   Try punching 4.5 into your calculator and squaring it. The answer is 20.25. That's a little high.

3. **If your result isn't quite right, try again, splitting the difference between your most recent guess and a lower number (if your guess was too high) or a higher number (if your guess was too low).**

   Split the difference between 4 and 4.5. Enter 4.25 on your calculator and square it. The answer is 18.0625, so you're getting closer.

4. **Repeat Step 3 until you get the most accurate answer you can.**

   Split the difference between 4.25 and 4.5. With 4.375, you get 19.140625, which is still just a little high. Try 4.3, 4.335, 4.355, 4.357, and 4.358. The last answer is 18.992. Close enough, though you can keep approximating if you like.

Square roots the effortless way

The effortless way to calculate a square root doesn't help your math skills, but it does get the job done. Pick one of two methods:

✔ Use your pocket calculator, scientific calculator, or smartphone. It may have a square root key:

$$\left(\sqrt{}\right)$$

✔ Use Microsoft Excel or Open Office Calc. The Excel square root function is =SQRT(*nnn*), where *nnn* is the number whose square root you want to find.

Example: Finding the Bytes On a Disk

You're installing disks with an advertised capacity of 320 gigabytes (GB). Your curiosity gets the better of you and you ask, "How many bytes is that?" You can solve this problem in two ways, and both are simple:

✔ **Method 1:** Take the known decimal number for 1 GB and multiply it by the number of bytes advertised. Table 11-2 earlier in this chapter tells you that known number is 1,073,741,824 bytes/gigabyte. One gigabyte is also 2^{30} bytes.

$$1,073,741,824\,\frac{\text{bytes}}{\text{gigabyte}} \times 320 \text{ gigabytes} = 343,597,383,680 \text{ bytes}$$

The answer is 343,597,383,680 bytes (although don't get too excited — the disk will have less capacity after it's formatted).

✔ **Method 2:** Take the known base 2 number for a gigabyte (also listed in Table 11-2) and multiply it by the number of gigabytes.

$$2^{30}\,\frac{\text{bytes}}{\text{gigabyte}} \times 320 \text{ gigabytes} = 320 \times 2^{30} \text{ bytes}$$

This answer is technically correct, but it probably doesn't satisfy your urge to see the answer in big base 10 digits. If you expand the factor of 2^{30} to become 1,073,741,824 bytes, you're repeating method 1.

Part III
Basic Algebra, Geometry, and Trigonometry

The 5th Wave   By Rich Tennant

"David's using algebra to calculate the tip.
Barbara— would you mind being a fractional
exponent?"

In this part . . .

*P*art III contains the "miracle math" chapters. Well, okay, the chapter about areas, perimeters, and volumes is only semimiraculous. But it's very handy and may be magical to some folks.

The big magic is in the power you get from the other chapters, those devoted to algebra (12), formulas (13), geometry (14), areas and volumes (15), and trigonometry (16). Those five concepts are powerful, and they usually inspire fear and awe when you first encounter them.

But consider this: Cave dwellers first feared fire because it could burn them. But when they mastered it, fire kept them warm and cooked their food. The modern equivalent is that when you master these subjects, your technical job will pay your heating bill and buy your groceries.

Chapter 12

Algebra and the Mystery of X

- -

In This Chapter

▶ Defining algebra and laughing at its "complexity"

▶ Boning up on algebra terms and operations

▶ Meeting *x*, the famous variable

▶ Testing out algebra in some real-world situations

- -

*A*lgebra is a branch of mathematics loved by a few and feared by many. Algebra is almost always in secondary education (high school) curriculums, but that doesn't mean everybody takes it. Some avoid it, and others who take it may come away scarred for life. Well, cut that out! You don't have to suffer from algebra-phobia. It's a simple part of math that gives you great control over your work.

The biggest concept is easy to grasp: in algebra, variables (which we cover in detail later in this chapter) are letters that represent numbers. That's it. And they only represent numbers until you solve a problem and replace the variables with numbers. As this chapter shows, you work on algebraic variables with the same math operations you use on numbers.

And why, you may well ask, should you care about these concepts? Because they're valuable in your everyday work. You may not even know when you're using algebra, but you can be sure pharmacy assistants, concrete contractors, computer techs, welders, cosmetologists, and roofers (to name just a few) calculate using algebra.

In this chapter, you get friendly with the names of the parts of an algebra statement. Then you go on to do simple (but essential) math operations.

Variables: Letters Represent Numbers, but the Math Is the Same

In arithmetic, you come across problems with numbers in them. In algebra, you come across problems with letters in them. For example:

$$a + 2 = b$$

The letter a represents an unknown number, but when you know a and add 2 to it, you can figure out what b is.

Before you start applying math to the variables, you need to know how to refer to the letters, numbers, and the expressions they create. The following sections get you up to speed on this vocab.

Understanding variables

A *variable* is a symbol that represents a value that can vary (hence the name *variable*). It stands for a number you don't know. The following are a few examples of variables:

$$a \; b \; c \; d \; x \; z$$

Variables can be any letter you want. Remember, a variable doesn't have a fixed or final value, at least not until you solve the problem. When you hear teachers, students, and co-workers talk about "solving for x," they're working with a variable (x) of unknown value. We cover variables in a great deal of detail later in this chapter.

Corralling constants

The opposite of a variable is a *constant,* which has a fixed value. The numbers 3, 2.5, ½, and π are constants, although constants can be numbers of any kind.

Sometimes a constant used in multiplication is called a *coefficient* or an *index.* Essentially, the words mean the same thing — a fixed multiplier. A constant can also be a fixed value in an expression or equation (see the next two sections).

There are many constants, and all of them are used in higher math, which is beyond the scope of this book. What we have here is something of a "short shot," or an attempt to provide a simple definition and a sample.

Examining expressions

An *expression* is a combination of symbols. That's it! An expression isn't necessarily equal to anything; it's like using a phrase rather than a whole sentence. Here's a sample expression:

$3 + 4 + 5$

This expression is made up entirely of constants. You can easily find an answer to what $3 + 4 + 5$ is, but that's not the goal here. What's important is how the symbols look when they're written out together.

But expressions don't have to consist only of constants. An *algebraic expression* also includes variables:

$a + b + c$

Here, you add some unknown quantity of something *(a)* to an unknown quantity of something else *(b)* and then add that total to another unknown quantity *(c)*.

For your purposes, the three symbols may represent sheetrock in three piles, pills in three containers, three time periods for cooking, or the quantities of three chemicals you need for coloring hair.

Getting a handle on equations

An *equation* is a combination of symbols, like an expression. The difference is that it has an equal sign (=) to show that two expressions are, well, equal. For example, here's one expression:

$3 + 4 + 7$

And here's another:

$5 + 8 + 1$

These numbers aren't much to write home about, but when you relate the expressions in an equation, the game heats up. For example, is the following true?

$3 + 4 + 7 = 5 + 8 + 1$

A little arithmetic on both sides of the equation gives you this:

$14 = 14$

So, where did algebra come from, anyway?

Here's something you may not know: the word *algebra* comes from the Arabic word *al-jabr*, and it means "restoration." But even though it's a word with Arabic roots, algebra goes back much farther to the Babylonians, a culture that sources say existed by 1728 BC, when Hammurabi, the great lawgiver, was king.

The Egyptians, Greeks, Indians, and Chinese made other math contributions, but algebra's ancient roots are in Babylonia.

It's the same Babylonia well known for archeological findings, the Bible, the story of the Tower of Babel, and the Hanging Gardens of Babylon. Babylon is thought to have been, at one time, the largest city in the world, so it's no surprise that these folks were brilliant masters of practical matters, such as

- ✔ All the construction and technical trades
- ✔ Technology, like metal working
- ✔ Astronomy and medicine
- ✔ Art and architecture

None of these pursuits would be possible without higher math — algebra.

The Babylonians were no slackers. They developed solutions to complex problems that today require elements of algebra that we don't cover in this book: linear equations, quadratic equations, and indeterminate linear equations.

The Babylonian system of math was *sexagesimal*, or a *base 60 numeral system*. Don't worry about what exactly that means; just know that it's the foundation of the modern systems of 60-second minutes, 60-minute hours, 360-degree circles, and so on.

One more thing: You may have heard of Omar Khayyám, the famous poet born in the Great Seljuq Empire (now part of Iran) in 1048 AD and partly known for the line "a Jug of Wine, a Loaf of bread — and Thou." But Omar Khayyám was famous during his times as a mathematician — he wrote the *Treatise on Demonstration of Problems of Algebra* in 1070.

What a relief! Now in the world of algebra, equations are far more interesting. For example:

$$a + b = 14$$

You don't know what a and b are yet, but the equation declares that they equal 14.

Taking time for terms

A *term* is any one part of an expression or equation (see the preceding two sections), separated from other terms by an addition (+) or subtraction (−) sign.

A term can be a constant, a variable, or the product of constants and variables. For example:

$$x + y + 3a - ab - xy + 3abxy$$

The terms are *x, y, 3a, ab, xy* and *3abxy.* The term *3a* is a combination of *3* and *a,* and it means "three times the amount of *a.*"

The term *xy* is a combination of *x* and *y,* and it means that *x* and *y* are multiplied together. Notice that you don't need a regular multiplication sign ($\times$), middle dot ($\cdot$), or asterisk (*) to show multiplication.

Variable Relationships: X and Her Friends

Although variables can be any letter, *x* is the most popular and enduring option for reasons beyond understanding. (Check out the nearby sidebar for a few theories.) The variable *x* has become the Queen of the Unknowns.

Other letters near the end of the alphabet, like *y* and *z,* are also popular variables. In some applications, the variables represent key words in the application. Here's an example from geometry (a topic we discuss in Chapter 14):

area of a rectangle = *lw*

In this example, the variable *l* stands for the *l*ength of the rectangle and the variable *w* stands for its *w*idth:

When you're comfortable with variables, you're ready to look at how they hook up with constants and group together with like variables, as the following sections discuss.

Best friends forever: The constant and the variable

Although variables and constants may seem opposed by definition, they're actually quite good algebraic buddies. In the real business of algebra, a constant and a variable are inevitably paired. They're best friends forever.

A constant and a variable together look like these examples:

$$a \quad 3b \quad 9c \quad \frac{1}{2}d$$

Look at the *3b* example. It represents the constant *3* multiplied by the variable *b*. This arrangement is just like multiplying in arithmetic, but with two exceptions.

- ✔ *b* is a letter.
- ✔ You don't need any sort of sign or symbol to show multiplication. It's understood.

You can represent *3b* in any of the following ways:

$$3 \times b \quad 3 \cdot b \quad 3 * b \quad b + b + b$$

The first three use multiplication symbols; the fourth expression adds *b* three times. But save yourself the effort — the expression *3b* says it all.

Tracking down the Queen of the Unknowns, and other variable fun

The origin of *x* as the go-to variable is, well, unknown. Speculation abounds, but nobody knows for sure. Here are a few possible answers:

- ✔ In the days of quill pens, *x* (and *y*, too) was the easiest to write.
- ✔ Dr. Ali Khounsary, Advanced Photon Source, Argonne National Laboratory, suggests that algebra solves for the unknown "thing" and that the word for *thing* or *object* in Arabic is *shei*, which was translated into Greek as *xei* and possibly shortened to *x*.
- ✔ Dr. Khounsary also points out, "It is also noteworthy that *xenos* is the Greek word for unknown, stranger, guest, or foreigner, and that might explain the reasons Europeans used the letter *x* to denote the 'unknown' in algebraic equations."

The use of *x* as a variable representing the unknown is not only widespread in math but has also permeated Western culture. For example:

- ✔ In 1985, Wilhelm Röntgen discovered a new type of radiation, which he temporarily called "X-rays" because they were unknown. Even though X-rays are called Röntgen radiation in some languages, in many other languages they're still known as X-rays.
- ✔ Science fiction novels, comics, movies, and television shows such as *The X-Files* use *x* to represent mystery and the unknown.
- ✔ Any pirate movie with a decent map shows a big *X* where the treasure is because "*X* marks the spot." (Well, okay, that may be more of a marking symbol than an unknown symbol.)

You may think that the *a* in the earlier example has no constant, but any solo variable has an imaginary *1* as a constant. Any number times 1 equals itself, and the same applies to variables multiplied by 1.

As we mention earlier in the chapter, you may come across the terms *coefficient* (such as the Coefficient of Friction or COF in automotive technology) or *index* (such as the *refractive index,* or how much light bends when it passes from air into a lens, if you work with optics) rather than constant. They all mean the same thing.

Simplifying variables: Variables of a feather flock together

In math, you can change the order of terms without changing the result. This property is called *commutativity,* and it's an important part of doing math operations with variables. For example, these two expressions are the same. The terms are just in a different order:

$$x + y = y + x$$

The expression $a + b + 3a$ has terms with both *a* and *b*. You can write the expression in a different order — $a + 3a + b$. This move doesn't change the result, but now you can more easily combine the *a* terms to get *4a*. You get the expression $4a + b$. We get formal about commutativity (and other fundamental math properties) in Chapter 13, but for now it's casual Friday.

Math Operations with Variables

Whether you're mixing liquids for hair color, assisting a pharmacist with compounding, or calculating the weight and coverage of mission tiles for a roof, you do math operations with variables. The following sections give you the skinny on performing the four main math operations (addition, subtraction, multiplication, and division) in equations with variables.

In algebra, as in arithmetic, an equation has two sides. These sides are called (get ready) the *left side* and the *right side.* For example:

$$a + b = c + d$$

In the example, the left side contains *a+b* and the right side contains *c+d*.

If an equation is *true,* you can do identical math operations to both sides of it without it becoming untrue.

In both math practice and in real life, algebra equations are true (assuming you set them up correctly) because you create them to determine unknown values (say, how many fan belts are in the stockroom) based on values you already know (for example, three fan belts on the top shelf and two identical belts on another shelf). Here's your fan belt equation:

$x = 3 + 2$

How about an untrue equation? For a false equation, try

$17 = 3 + 2$

That's just plain bad arithmetic. Here's a more creative false algebraic equation:

$b = 2b + 3b$

How can three of something (*3b*) and two of something (*2b*) add up to be one of something (*b*)? Unless *b* just happens to be zero, this equation is mathematical nonsense.

When an algebra equation is true, you can add, subtract, multiply, or divide both sides of an equation (although you can't divide by zero).

But wait! There's more! (We sound like a late-night infomercial.) Just like numbers, you can add, subtract, multiply, and divide variables. These operations are how you *reduce* equations to their simplest terms; the following sections give you more information on each operation.

 The better you can do math operations on variables, the faster and easier your work becomes. All the techniques in this section apply in real-life scenarios, mostly in a simpler form than they appear here. Real-life situations are what show up in story problems, which we show later in the chapter.

Adding variables

The most important rule in adding variables is that you may only add like terms. Say you're baking apple pies and the recipe calls for six pounds of apples. You have only four pounds on hand, but you remember that you also have two pounds of oranges. Don't do it! Apples and oranges don't mix in apple pies, and neither do different variables in algebraic expressions.

To keep from ruining your algebra pie, follow these rules:

- ✔ You can add terms with the same variable.
- ✔ You can't add terms with different variables.
- ✔ The order of the terms doesn't matter, and you can rearrange the order.

The following expression has six terms, separated by addition signs. It has three different variables *(a, b,* and *c)*. You need to treat each variable separately.

$$3a + 2b + 4c + 2a + b + 4c$$

To add terms in an expression, first identify the like variables:

- ✔ The first variable is *a:*

    $$3\mathbf{a} + 2b + 4c + 2\mathbf{a} + b + 4c$$

- ✔ The second variable is *b:*

    $$3a + 2\mathbf{b} + 4c + 2a + \mathbf{b} + 4c$$

- ✔ The third variable is *c:*

    $$3a + 2b + 4\mathbf{c} + 2a + b + 4\mathbf{c}$$

Now that you know the like variables, your next operation is to combine them:

1. **Starting with the variable *a*, use the commutativity property of addition to rewrite the expression.**

    $$3\mathbf{a} + 2b + 4c + 2\mathbf{a} + b + 4c$$

    Check out "Variables of a feather flock together" earlier in the chapter for more on this property. Now the *a* terms are together:

    $$3\mathbf{a} + 2\mathbf{a} + 2b + 4c + b + 4c$$

2. **Add the coefficients of each of the "a" terms.**

    You end up with 5a:

    $$3\mathbf{a} + 2\mathbf{a} = 5\mathbf{a}$$

3. **Rewrite expression with simplified variable *a*.**

    After you insert the newly condensed *5a* into the expression, the expression looks like this:

    $$5\mathbf{a} + 2b + 4c + b + 4c$$

4. **Repeat Steps 1 through 3 for the variables *b* and *c*.**

   You must continue to combine the like terms for all remaining variables until all are in their simplest forms. The single variable *b* has an invisible constant of 1.

5. **Rewrite the expression, with all the reduced terms, in its simplest form.**

   Your final expression looks like this:

   $5a + 3b + 8c$

Subtracting variables

Get ready to tackle that pesky subtraction sign. You don't have to worry about any new rules or properties being thrown unexpectedly your way. The only thing that may give you a little problem is remembering to place the operation signs in the proper order.

Approach the subtraction signs with caution, to ensure you get the expression simplified correctly.

The following example shows you how to move the operation signs correctly. Start with an expression such as the following:

$3a + 2 - 2a - 1$

Rewrite the expression to combine the variables and constants. Pay particularly close attention to the subtraction sign preceding the variables (that's *−2a* in the example) you're working with. You need to make sure that it stays "attached" to the correct variable. Your rearranged expression looks like this:

$3a - 2a + 2 - 1$

Now, you're ready to combine the like *a* and number terms (separately, of course):

$3a - 2a = 1a$

$2 - 1 = 1$

Together, the result is

$a + 1$

The expression is now in its simplest form, and you can go no farther.

Multiplying variables

In order to multiply variables, you need to be familiar with the rules for multiplying exponents (which we discuss in Chapter 11.) The rules for multiplying exponents are similar to those for adding and subtracting variables; see the preceding two sections.

When you want to multiply variables, keep these rules in mind:

✔ You can multiply terms that have the same variable.

✔ You can't multiply terms with different variables.

✔ You multiply expressions by adding the exponents and keeping the same base (variable).

✔ The order of a term's parts doesn't matter; you can rearrange the order and separate the parts.

A term with no indicated exponent has an invisible exponent of 1. Any item exponentiated (raised) to the power of 1 is itself. For example, 2^1 is 2 and a^1 is a.

Multiplying simple variable terms

The most common uses of multiplying exponents in various vocations are square measure and cubic measure. In square measure, familiar to roofers and landscapers, the factors have a power of 1 and the answer has the power of 2. For example, length in feet multiplied by width in feet equals area in square feet. Cement masons and landscapers will make volume calculations, where one factor has a power of 1 and the other has a power of 2. For example, area of a patio in square feet multiplied by the thickness of the concrete in feet equals volume in cubic feet (the power of 3). In the following example, both terms have the same base, so you just add the exponents m and n together and rewrite the result with the same base.

$$(a^m)(a^n) = a^{(m+n)}$$

This example uses the numbers *2* and *3* for exponents in place of m and n.

$$(a^2)(a^3) = a^{(2+3)} = a^5$$

It's the same story. Add the exponents. That's multiplying in variable land.

Dealing with more complex multiplication

Fortunately, everyday work in the trades doesn't require computation above cubic measure (the power of 3). However, your education to improve yourself or get into a trade may include complex multiplication.

The following example is an expression with multiple variables and multiple constants:

$$(3a^4b^7c^{12})(-5a^9b^3c^4)$$

This example has constants and three variables. Now, reduce the expression to its simplest form, a form where the expression can no longer be modified or reduced further. Here's how it works:

1. **Take each of the variables and separate them from each other.**

   Doing so makes regrouping them easier. Here's what you end up with:

   $$(3)(a^4)(b^7)(c^{12})(-5)(a^9)(b^3)(c^4)$$

   Notice that this spread is all one term. It has a lot of constants and variables but no plus or minus signs separating anything into multiple terms. (The "−" in "−5" isn't functioning as a subtraction sign — it's a negative sign. It's in parentheses, and that means it's attached to the 5.)

2. **Regroup the like variables.**

   You're not doing any multiplying yet; you're just matching everybody up to make the multiplication easier. In this case, you have four separate pieces to the equation:

   $$(3)(-5) \quad (a^4)(a^9) \quad (b^7)(b^3) \quad (c^{12})(c^4)$$

3. **Do the math by multiplying the constants and adding the exponents of the variables.**

   Of course, you're calculating each variable separately:

   $$(3)(-5) = -15$$

   $$a^{4+9} = a^{13}$$

   $$b^{7+3} = b^{10}$$

   $$c^{12+4} = c^{16}$$

4. **Reunite all the terms by rewriting the constants and variables as one expression.**

   Just put the pieces back together:

   $$-15a^{13}b^{10}c^{16}$$

This example is more complex than what you're likely to find in your everyday work. But now you know how to tackle it if it does come up.

Dividing variables

Dividing variables is similar to multiplying variables. The difference is that with division, you subtract the exponents instead of adding them. (Check out the preceding section for more on multiplying variables.)

In order to divide variables, make sure you're familiar with the rules for dividing exponents, which we cover in Chapter 11. The rules are similar to those for adding, subtracting, and multiplying variables we present in the preceding three sections.

When you're dividing variables

✔ You can divide expressions that have the same variable.

✔ You can't divide expressions with different variables.

✔ You divide expressions by subtracting the exponents.

✔ You can rearrange the order of the terms and separate the parts because the order doesn't matter.

Be familiar with the difference between the numerator and the denominator of a fraction — variable division is written in a form that looks very similar to a fraction. We discuss fractions more in Chapter 8, but for now, remember that the *numerator* is the top number in a fraction and the *denominator* is the bottom number.

Tackling simple variable division

In the following example, you divide 6*a* by 2*a*. Stack the numerator over the denominator and reduce the fraction to its lowest possible terms.

$$\frac{6a}{2a} = \frac{6}{2} \times \frac{a}{a} = 3$$

 This kind of variable division occurs every day in the trades.

The variable in the denominator can't be equal to zero. It's impossible to divide by zero.

When the variable terms have exponents, you just subtract the denominator's exponent from the numerator's exponent. Because both terms in this example have the same base, you subtract the exponent *n* in the denominator from *m* in the numerator.

$$\frac{a^m}{a^n} = a^{(m-n)}$$

This example uses the numbers 5 and 3 for exponents rather than m and n.

$$\frac{a^5}{a^3} = a^{(5-3)} = a^2$$

Here's a simple illustration of what this principle does. You can break the expression down into separate division problems:

$$\frac{a^5}{a^3} = \frac{a \times a \times a \times a \times a}{a \times a \times a} = \frac{a}{a} \times \frac{a}{a} \times \frac{a}{a} \times \frac{a}{1} \times \frac{a}{1}$$

Now, watch this trick carefully. Because

$$\frac{a}{a} = 1$$

and

$$\frac{a}{1} = a$$

the expression simplifies to

$$1 \times 1 \times 1 \times a \times a \text{ or } a \times a \text{ or } a^2$$

Same results, but a lot more work than simply subtracting the exponents.

Diving into more complicated variable division

You don't frequently encounter division of multiple variables in everyday work, but such problems can theoretically come up in surface calculations (for example, roof tiling or solar panel installation). This example contains both multiple constants and multiple variables:

$$-\frac{3a^9 b^7 c^{12}}{5a^4 b^3 c^4}$$

Although the expression has several variables, the process for reducing it is relatively simple.

1. **Separate the constants and like variables, watching out for any attached signs.**

   In this example, remember to keep the negative sign with the ⅗ fraction.

   $$-\frac{3a^9 b^7 c^{12}}{5a^4 b^3 c^4} = -\frac{3}{5} \times \frac{a^9}{a^4} \times \frac{b^7}{b^3} \times \frac{c^{12}}{c^4}$$

The expression breaks down into separate parts, but the mathematical meaning is the same.

TIP

The expression doesn't actually contain any multiplication signs, but using them helps to separate each of the variable terms. Just remember that your main goal here is division.

2. **Divide each variable by subtracting the denominator exponent from the numerator exponent.**

   We show you the *a* variable here:

   $$\frac{a^9}{a^4} = a^{(9-4)} = a^5$$

   After you divide all three variables, they're in their simplest forms.

3. **Rewrite the expression.**

   You have now simplified the expression to its lowest form.

   $$-\frac{3}{5}a^5 b^4 c^8$$

Example: How Many Oranges Are in that Orange Juice?

A chapter without real-life problems is like a day without sunshine. Here's some sunshine:

REAL WORLD EXAMPLE

You're working in an upscale hotel, and the restaurant has a novel policy of making orange juice from (get this) freshly squeezed oranges. Here's what you need to know:

✔ According to the one source, an orange has about 2 ounces of juice. You need 3 to 4 medium oranges to make an 8-ounce glass of juice, so for this example, assume it takes 4 oranges.

✔ An orange weighs about 9 to 11 ounces. For this example, assume 10 ounces.

✔ 1 U.S. gallon = 128 U.S. fluid ounces.

Your boss wants you to make one U.S. gallon of orange juice. How many oranges do you need?

First, evaluate the information you have. The number of 8-ounce glasses involved is irrelevant here. Life problems sometimes have extra information you don't need. Also, an orange's weight doesn't really matter either, unless you need to go buy a quantity of oranges sold by weight.

What *does* matter is how many oranges (producing 2 ounces of juice each) you need to fill a 128-ounce gallon container. Follow this process:

1. **Let x be the number of oranges you need.**

   Now you've got one term in your equation. Not much, but it's a start:

   x oranges =

2. **Develop a constant showing the number of fluid ounces of juice per orange.**

   You get 2 fluid ounces per orange, so your next term is

   $$\frac{2 \text{ fluid ounces}}{1 \text{ orange}}$$

3. **Put the two terms together.**

   Now you've got the whole left side of the equation:

   $$x \text{ oranges} \times \frac{2 \text{ fluid ounces}}{1 \text{ orange}} =$$

4. **Develop an expression that represents your desired result and place it on the right side of the equation you started in Step 3.**

   You know the desired result here is 128 fluid ounces, so plop that down on the right-hand side of the equal sign to create the full equation:

   $$x \text{ oranges} \times \frac{2 \text{ fluid ounces}}{1 \text{ orange}} = 128 \text{ fluid ounces}$$

5. **Eliminate the units and simplify the equation.**

   What remains is

   $x \times 2 = 128$ or $2x = 128$

6. **Reduce the equation.**

   In this case, divide both sides by 2 to come up with your final answer:

   $x = 64$

   You need 64 oranges producing about 2 fluid ounces of juice each to make up a 128-fluid-ounce gallon.

Converting oranges into pounds of oranges

Unfortunately, the produce supplier doesn't deliver "by the orange." The supplier sells oranges by the pound. Because you need 64 oranges and oranges weigh 10 ounces, calculate how many pounds of oranges you need.

1. **Let *x* be the number of pounds of oranges you need.**

   x pounds =

2. **Determine how many ounces 64 oranges will be when each orange weighs about 10 ounces.**

   64 oranges × 10 ounces/orange = 640 ounces

3. **Develop a constant showing the relationship between ounces and pounds.**

   $$\frac{1 \text{ pound}}{16 \text{ ounces}}$$

4. **Combine ounces from Step 2 with the constant from Step 3.**

   $$640 \text{ ounces} \times \frac{1 \text{ pound}}{16 \text{ ounces}} = \frac{640 \text{ pounds}}{16} = 40 \text{ pounds}$$

   The answer is 40 pounds.

Example: Medications In the Pillbox

As a certified nursing assistant (CNA), you need to place doctor-prescribed blood pressure medications in a container that holds the patient's next seven days of medications. The blood pressure meds are

- ✔ Accupril: 40 mg tablet once daily
- ✔ Amlodipine: 5 mg capsule once daily
- ✔ Dyazide: 37.5-25 mg tablet once daily

How many pills of each kind do you need to deposit for each day? Because you're filling a seven-day container, how many pills of each kind do you give this patient each week?

This question may seem like a trick because of its simplicity, but it helps demonstrate problem solving techniques. You can solve the first part of the problem, how many pills of each kind you need each day, by recognizing that one pill of each kind is required each day. In effect, the problem gives you the answer. This simple preliminary process is called *inspection*. The answer is that one pill of each kind is required per day. You get this from the "once daily" wording of the problem.

You can solve the second part of the problem, how many pills of each kind are required each week, in your head, but you can also represent it as an algebraic expression. Here's how to set it up:

1. **Let x be the number of Accupril tablets you need in a week and let a be the number of Accupril tablets provided in a day.**

   $x =$

2. **Build a simple equation.**

   Because you know that Accupril is dispensed every day and that there are 7 days in a week, you can create the following equation:

   $x = a + a + a + a + a + a + a$

3. **Add the a variables and substitute the result in the equation.**

   That sequence looks like this:

   $a + a + a + a + a + a + a = 7a$

   $x = 7a$

4. **Solve for x.**

   This process is pretty simple because you know from your inspection that the number Accupril tablets you dispense per day is one.

   $x = 7a = 7 \times 1 = 7$

It takes 7 Accupril tablets to fill the container for a week. You can use similar equations to determine how many Amlodipine and Dyazide pills to give out weekly. (Actually, because you're giving out the same number of all three pills daily, you can use the exact same equation, substituting one of the other medicines as the value of x.)

The answers get far more interesting when the distribution of pills throughout the week is uneven, or when achieving the correct dosage takes multiple pills.

Chapter 13

Formulas (Secret and Otherwise)

$\cdots\cdots\cdots\cdots\cdots\cdots\cdots\cdots\cdots\cdots\cdots\cdots\cdots\cdots\cdots\cdots$

In This Chapter

▶ Understanding the basics of formulas and their properties

▶ Converting and managing units in formulas

▶ Turning formulas you know into other helpful formulas

$\cdots\cdots\cdots\cdots\cdots\cdots\cdots\cdots\cdots\cdots\cdots\cdots\cdots\cdots\cdots\cdots$

The word *formula* has a special meaning in mathematics, but it's also everywhere in pop culture. It can mean many things, but all the uses of the word come from the same idea: It's method of doing something that should produce the same result every time. In mathematics, a *formula* is a compact rule for getting something done mathematically. It's the math equivalent of "Never run with scissors" or "Always say please and thank you." It always works.

What's important to you is that, no matter what career you're in, you use formulas to speed up math. You can use off-the-shelf formulas (and we offer many in this book), but from time to time you make your own. And your own custom formulas make your work go even better because they apply directly to what you're doing.

In this chapter, you get the formula for working with formulas. You discover what the parts of a formula are and what alterations you can make to its structure by moving parts around. You also work with units in formulas and whip up your own home-brewed formulas.

Formulas in the world outside math

In addition to the math meaning, a formula can be a recipe created by a scientist, chemist, pharmacist, or chef for making something. One important kind of formula is infant formula, invented about 1867, which is supposed to contain nutrients to substitute for a mother's breast milk. Why it's referred to as formula is a mystery.

The most famous scientific formula is Einstein's $E = mc^2$. Although you know it best for its role in nuclear reactions, it's also a part of the math in particle physics and the physics of the Big Bang and black holes.

A formula can be a trade secret. The most famous secret formula is for Coca-Cola, and the company makes a big deal about how it's locked away in a vault or about how few people know it. A different opinion is that the "secret" is more part of a great marketing campaign than anything else.

In movies, books, and television, a formula is "the same old plot," predictable and uncontroversial. That's usually what you can expect from situation comedies on TV (as well as, interestingly enough, the 1980 movie *The Formula*, a thriller about finding a formula for artificial gasoline).

Following the Formula for Building a Formula

The word *formula* comes from the Latin word *forma; formula* means "little rule" or "little method." A formula is usually expressed as an equation, so it contains an equal (=) sign. That means something (a desired result) is equal to something else (the factors you need to get the result).

A formula uses *variables* (letters) and may have *constants* (numbers). For example, you express the area of a rectangle as $A = L \times W$, where A is the area, L is the length, and W is the width. The length and the width are variables, and they're represented by letters. They can be any number. They can vary, and that's why they're called variables.

A simple example of a formula with a constant is the one that shows the relationship of the radius of a circle to the diameter. That formula is $d = 2r$. You say this formula as "the diameter is equal to two times the radius." The radius may vary (being a variable), but the *2* is constant (being an unchanging number).

Formulas have three properties you can use to manipulate them. These properties are 100 percent reliable, worked out and proven by professionals a long time ago in a galaxy far away. The properties are *associativity, commutativity,* and *distributivity*. And no, we're not making these words up as we go along — honest.

Property A: Associativity

Join the association. In a formula, how you group the terms doesn't matter as long as the sequence doesn't change. Rearranging parentheses doesn't affect the value of an expression. This property works for addition and multiplication. In this example, adding 1, 2, and 3, you can write the terms two ways:

$$(1 + 2) + 3$$
$$1 + (2 + 3)$$

The *1* and *2* are associated. You add 1 and 2 first. Then add the result, 3, to 3 to get 6. But you can also associate 2 and 3. You add 2 and 3 first. Then add the result, 5, to 1 to get 6.

In multiplication, you may have a similar expression:

$$(1 \times 2) \times 3$$

Same as the addition associativity, you can also multiply

$$1 \times (2 \times 3)$$

What about subtraction and division? Nope. The property doesn't apply to subtraction or division. If you try it, it's not guaranteed to work.

Property C: Commutativity

Commutativity isn't an activity you do on your commute. It means that changing the order of the elements in an equation doesn't change the result.

For example, you can use one of two formulas for putting on your shoes:

Putting shoes on = putting on left shoe + putting on right shoe

Putting shoes on = putting on right shoe + putting on left shoe

It doesn't matter which shoe you put on first. The end result is the same. The same is true with addition:

$$1 + 2 + 3 = 6$$
$$3 + 2 + 1 = 6$$
$$2 + 1 + 3 = 6$$

No matter how you rearrange the terms, the answer is 6. And like its cousin associativity (see the preceding section), commutativity works with multiplication, too.

$$4 \times 5 \times 6 = 120$$
$$6 \times 5 \times 4 = 120$$
$$5 \times 4 \times 6 = 120$$

No matter how you rearrange the terms, the answer is 120.

Commutativity applies to addition and multiplication. The property doesn't apply to subtraction or division, and it's not guaranteed to work with those operations.

Property D: Distributivity

Distributivity is one more key concept in working with equations. When you multiply one term inside parentheses by another term outside the parentheses, you can (and must) distribute the multiplication to each of the terms inside the parentheses. Distributivity is a common way to break down formulas for solutions — making the complex become simple.

The following example uses numbers. Here, you multiply a term inside parentheses (4 + 2) by a term outside the parentheses.

$$3 \times (4 + 2)$$

Say this expression as "three times the quantity four plus two." The conventional approach is that you calculate what's inside the parentheses first.

$$3 \times (6) = 18$$

The answer is 18. But you can also distribute the multiplication to each of the terms inside the parentheses.

$$3 \times (4 + 2) = 3 \times (4) + 3 \times (2)$$
$$3 \times (4) + 3 \times (2) = 12 + 6 = 18$$

Again, the answer is 18. It doesn't seem dramatic at this point, but this concept is a magic property — like Aladdin's beat-up old lamp or Jack's beans — when you need to break down complex equations into simpler terms.

Distributivity is also what you use when the items inside the parentheses aren't the same and you can't combine them.

In the following example, the crew wants burgers and fries for lunch. There are six co-workers, and each one wants one burger and two orders of fries. So how many of each item do you buy? You can use following spoken formula, but it may be misleading:

> The total order is that six workers each get one burger and two orders of fries.

Does that mean that each worker gets one burger but you only pick up two orders of fries for everybody? Or does it mean that each worker gets a burger and each gets two orders of fries? To clear up confusion, write the order as a formula.

> Total order = 6 workers × (1 burger + 2 fries)

Now you see two things. First, each worker wants one burger and each worker wants two orders of fries. Second, you see that you can't add the items inside the parentheses. Burgers and fries are different units.

So you figure the total order by using the distributivity property.

> Total order = (6 × 1 burger) + (6 × 2 fries)

> Total order = (6 burgers) + (12 fries)

Now you can easily order 6 burgers and 12 fries at McFastfood's. And if you know the item prices, you can quickly apply some money math to the two quantities to calculate the total cost of the order.

Working from a Formula to a Solution

An *equation* is a mathematical statement. It has two expressions and an equal sign. One expression is at the left of the equal sign, and the other is at the right. So the assertion is that the left side is equal to the right side.

Equations have variables (represented by letters) and constants (represented by numbers). If that sounds like the definition of *formula* from the preceding section, that's because it is. Formulas are equations.

Here is an equation made up of numbers.

> 5 = 2 + 3

Here is an equation made up of variables. It's the formula for the area of a rectangle.

$$A = L \times W$$

In the equation for area, A represents the area to be calculated, L represents the length, and W represents the width.

Most equations are true, but you may occasionally find a false one:

$$2 + 2 = 5$$

Formulas are always equations. When you know what's in an equation and what you can do with it, you can then modify any formula to solve problems or adapt it to meet your special needs. The following sections show you how to do just that with the basic math operations (check out Chapters 4 and 5 for more on those), as well as how to apply a special multiplication rule.

Applying the same operation on both sides of the equal sign

When an equation is true (and they generally are), you do math operations on it to make it simpler:

- ✔ Add the same quantity to both sides.
- ✔ Subtract the same quantity from both sides.
- ✔ Multiply both sides by the same quantity.
- ✔ Divide both sides by the same non-zero quantity.

Don't divide by zero. It's not allowed, and the results don't make any sense anyway!

The equation still holds true as long as you do the same operation to both sides.

Adding it up

In this sample math addition equation, both sides are equal to five.

$$5 = 2 + 3$$

Now add 4 to each side.

$$5 + 4 = 2 + 3 + 4$$
$$9 = 9$$

Because you performed the same action on both sides, the sides are still equal. The same is true when you work with symbols. If you're figuring the area of a floor for carpet or flooring, or a surface area for a patio, you may want to add some additional square feet just to be on the safe side. The formula for the area of a rectangle is

$$A = L \times W$$

Say you want to give yourself 12 square feet of leeway on the project. Add 12 square feet to the answer even before you know the answer to a specific area problem.

$$A + 12 = (L \times W) + 12$$

Both sides increase without affecting the answer. In real life, the area *(A)* is unknown, so you can actually just make the fudge factor part of the equation.

Joe's area $A = (L \times W) + 12$

Now you have a custom formula. It's based on the standard formula, plus your allowance for extra square footage. As long as you understand why the extra square footage is there, you won't have any problem.

What about subtraction?

The principle covered in the preceding section also works for subtraction. Both sides of this equation are equal to 2:

$$2 = 5 - 3$$

Now subtract 1 from each side.

$$2 - 1 = (5 - 3) - 1$$
$$1 = (2) - 1$$
$$1 = 1$$

Both sides are equal to 1.

Mixing in multiplication

You're a manager at a concrete company. You know that each of your in-transit mixers (or cement trucks, as they're popularly known) holds 8 cubic yards. You need enough ready-mix concrete to fill 12 trucks on Monday morning. How much concrete is that?

The formula is simple, as some of the best formulas are. First start with the basic version:

> 1 truck = 8 cubic yards

Multiply both sides of the equation by the same amount (12).

> $12 \times (1 \text{ truck}) = 12 \times (8 \text{ cubic yards})$
>
> 12 trucks = 96 cubic yards

Deceptively simple, eh? But it works. And it doesn't just work for concrete. Multiplying both sides of the equation by the same amount is the basis for all recipe scaling. Scaling is most visible in culinary arts, especially in large food service establishments. It applies to sauces, beverages, pastries, and even the number of steaks required. The same scaling is applicable in the laboratory and ordering parts for manufacturing production runs.

Conquering the great divide

You divide both sides of an equation by the same amount to make it simpler. Just don't divide by zero. The following kitchen management example (pies and slices), is the opposite of the concrete example. Just make sure your pies aren't as hard as concrete.

You supervise food preparation in a retirement home. The boss lady tells you that you can get 288 slices of pie from 36 pies. How many slices should you cut each pie into?

1. **Start with the formula your manager gave you.**

   That's 288 slices = 36 pies

   That's interesting, but not useful — you're looking for the number of slices per pie.

2. **Divide both sides of the equation by 36 to get the slices in one pie.**

   $$\frac{288 \text{ slices}}{36} = \frac{36 \text{ pies}}{36}$$

   8 slices = 1 pie

   The answer is 8 slices per pie.

You first saw the division sign (÷) in elementary school, but did you know that it has a name? It's an *obelus,* and the plural is *obeli.*

Converting units with a special multiplication rule

Sometimes, the units in your formula may not match the units you use to order supplies. In this situation, you need a *unit conversion factor* to translate the units you have into the units you need. Luckily, using a conversion factor is the same as multiplying both sides of an equation by 1.

The trick here is that 1 can take many forms. Here's how it works:

1. **Start with the conversion formula.**

   Say you're a cement mason calculating a pour. You measure the pour in cubic feet, but you have to order the concrete in cubic yards. You start by noting how many cubic feet are in a cubic yard:

   1 cubic yard = 27 cubic feet

2. **Place the term with the units you have over the term with the units you need in a fraction, keeping the units intact.**

   Because the two terms (1 cubic yard and 27 cubic feet) are equal (or *equivalent*), you can place one over the other in a fraction. Because 1 cubic yard is equal to 27 cubic feet, this fraction is essentially the same as dividing 27 cubic feet by 27 cubic feet, which equals 1, as the following equation shows.

   Hang onto the units! The math won't make sense without them. After all,

$$\frac{1}{27} \text{ doesn't equal 1, but } \frac{1 \text{ cubic yard}}{27 \text{ cubic feet}} \text{ sure does.}$$

$$\frac{1 \text{ cubic yard}}{27 \text{ cubic feet}} = \frac{27 \text{ cubic feet}}{27 \text{ cubic feet}} = 1$$

$$\frac{1 \text{ cubic yard}}{27 \text{ cubic feet}} = 1$$

   Congratulations! You just made a conversion factor. Each side of the equation is equal to 1.

3. **To use your conversion factor, figure out how much of a product you need in one unit and then create a simple conversion formula.**

   If you're pouring an 8 foot x 20 foot slab 4 inches (.333 feet) thick, you figure out the cubic feet of concrete needed. That's 53.28 cubic feet ($8 \times 20 \times .333$). Then you calculate the concrete order with this simple formula:

   V cubic yards = 53.28 cubic feet

4. **Multiply both sides by 1.**

On the left side, use a real 1. On the right side, use the conversion factor from Step 2, which is equal to 1.

$$V \text{ cubic yard} \times 1 = 53.28 \text{ cubic feet} \times \frac{1 \text{ cubic yard}}{27 \text{ cubic feet}}$$

5. **Simplify the equation.**

Notice that the cubic feet on the right cancel out because that unit appears in both the top and bottom of the right side of the equation.

$$V \text{ cubic yards} = \frac{53.28}{27} \text{ cubic yards} = 1.973 \text{ cubic yards}$$

The answer is 1.973 cubic yards.

Calculating Speed, Time, and Distance: Three Results from One Formula

Everybody loves bargains. Imagine this one: Buy one and get two for free! That's the way lots of formulas work. Many formulas have three parts, and when you know how to solve for one of the parts, you can reconstruct the formulas to solve for the other parts. This bonus is true of several area calculations. It's especially true when you convert units — length, area, volume, and weight. (See the preceding section for more on converting units.)

A really obvious example is when you calculate speed, time, and distance. If your work requires travel, especially driving, you have to deal with these questions:

✔ How fast do I have to go to get there on time?

✔ How long will it take to drive there?

✔ If I drive this fast, how much distance will I cover?

You can also find speed/time/distance calculators on the Internet. But where's the fun in that?

Solving for speed

Speed is your rate of motion, and you express it in units of distance per unit of time — miles per hour, feet per second, and so forth. It's the rate at which you're covering distance. The formula for speed is

$$v = \frac{d}{t}$$

In the formula, v stands for velocity (speed), d stands for distance, and t stands for time. Now that should be easy to remember.

To get formal about it, speed is a *scalar* quantity, while velocity is a *vector* quantity. *Velocity* is the rate of change of position, which means it has direction. *Speed* is direction-independent and is just the magnitude of the velocity vector. You express both in the same units (miles per hour, kilometers per hour, feet per second, meters per second, and so forth). As a nonmathematician, you use the terms interchangeably.

Here's one of those "solve it in your head" examples. If you drive 55 miles in one hour, what's your speed? Put your data into the formula.

$$v = \frac{55 \text{ miles}}{1 \text{ hour}}$$

Separate the units from the number.

$$v = \frac{55 \text{ miles}}{1 \text{ hour}} = \frac{55}{1} \frac{\text{miles}}{\text{hour}}$$

Your speed is 55 miles per hour. This calculation gets more complex, of course, when the time is an odd amount (for example, 2.46 hours) or appears as non-hourly units (37 minutes). You may have to do some conversion, but after you do, the formula works fine.

Solving for speed and its cousins, time and distance, is a frequent component of technical work. The worlds of time/distance/speed, money, and time expended may seem secondary, but in professional work, they're always important and always there.

Solving for time

When you know the formula for calculating speed (covered in the preceding section), you can easily convert it into a valuable formula where the unknown is time. Just follow these easy steps:

1. **Start with what you know, the formula for speed.**

   $$v = \frac{d}{t}$$

2. **Multiply both sides of the equation by *t*.**

   $$v \times t = \frac{d}{t} \times t$$

   $$vt = d$$

3. **Divide both sides of the equation by *v*.**

   $$\frac{vt}{v} = \frac{d}{v}$$

   $$t = \frac{d}{v}$$

The formula for finding time (when you know distance and speed) is distance divided by speed.

Solving for distance

But wait! There's more! What they say in the infomercials applies to the calculation of time/distance/speed. In addition to the formulas in the preceding sections, one more formula makes the time/distance/speed formula handier than a Ginsu knife. Here's how you develop the formula for distance.

1. **Start with what you know, the formula for speed.**

   $$v = \frac{d}{t}$$

2. **Multiply both sides of the equation by *t*.**

   $$v \times t = \frac{d}{t} \times t$$

   $$vt = d$$

3. **Write the equation with *d* on the left and the factors *v* and *t* on the right.**

   $$d = vt$$

   Distance is equal to speed multiplied by time.

A quick conversion: If you drive at 60 miles per hour (where legal, of course), you travel 60 miles in 1 hour or 1 mile in 1 minute. At this speed, you can easily figure that you'll drive (for example) 35 miles in 35 minutes.

That's all you need to do. One formula produces three formulas for you. When you know the formula for speed, you get the time for nothing and the distance is free (apologies to Dire Straits here).

Example: Cement Masonry — Pouring City Sidewalks

You're a cement contractor in Riverside, Illinois, about 15 miles west of Chicago. Your company gets a lot of work from the city for its sidewalk replacement program.

Sidewalks are compliant with the Americans with Disabilities Act (ADA) and are 60 inches wide. Although sidewalks must be thicker where they cross driveways, you're interested in the majority of sidewalks on public land, which must be 5 inches thick.

How many yards of concrete do you need for a 50 foot run? Based on the results of that calculation, what's the approximate amount of concrete you need for each running foot of sidewalk? Develop a custom formula for each calculation.

1. **Start with the formula you know for figuring volume of a rectangular space.**

   $$V = L \times W \times H$$

   The volume *(V)* is equal to the length *(L)* multiplied by the width *(W)* multiplied by the height *(H)*. The height of a sidewalk is its thickness.

2. **Convert the units to feet.**

   You know the length you want in feet — 50 feet. You know the width and height, but those dimensions are in inches, so you need to use a conversion factor.

   $$\text{Width in feet} = 60 \text{ inches} \times \frac{1 \text{ foot}}{12 \text{ inches}} = \frac{60}{12} \text{ feet} = 5 \text{ feet}$$

   $$\text{Height in feet} = 5 \text{ inches} \times \frac{1 \text{ foot}}{12 \text{ inches}} = \frac{5}{12} \text{ foot} = .417 \text{ foot}$$

   Watch your units. Otherwise, you may get a lot more concrete than you ever dreamed of delivered to the building site.

   The width of the sidewalk is 5 feet, the thickness is .417 feet, and the length is 50 feet. Now you're ready to calculate.

3. **Insert your information in the formula and calculate.**

   V in cubic feet = 50 feet × 5 feet × .417 feet

   V in cubic feet = 104.25 cubic feet

   The answer, in cubic feet, is 104.25. But you want to know how many cubic yards to order.

4. **To convert to cubic yards, use the cubic feet to cubic yard conversion factor from earlier in this chapter.**

   $$\frac{1 \text{ cubic yard}}{27 \text{ cubic feet}} = 1$$

   V in cubic yards = 104.25 cubic feet × $\frac{1 \text{ cubic yard}}{27 \text{ cubic feet}}$

   V in cubic yards = $\frac{104.25}{27}$ cubic yards = 3.86 cubic yards

   The answer is 3.86 cubic yards (or "yards," as they say) to pour 50 feet of sidewalk 5 feet wide and 5 inches thick.

5. **To figure out how much concrete you need for 1 running foot of concrete, just divide your Step 4 total by 50.**

   V for 1 running foot = $\frac{3.86 \text{ cubic yards}}{50 \text{ running feet}} = \frac{0.0772 \text{ cubic yards}}{1 \text{ running foot}}$

   The answer is 0.0772 cubic yards per running foot. Congratulations! You now have you own custom formula or conversion factor. If, for example, someone wants to know how much concrete to order for a 22 foot run, just multiply 22 by your factor of 0.0772. (By the way, that answer is 1.698 cubic yards.)

Example: Lunch Time — Buying Burgers and Fries

Man (and woman) doth not live by work alone. Nor do they live by math alone. Sometimes, they must eat lunch.

Whether you're on a construction crew, working in a lab, or coaching youth soccer, eventually you get tagged to go get the food (unless you work in the culinary arts; you can probably eat the food right where you work).

Okay, so you're the lucky one to go to Burger Duke to buy the lunch. You have three kinds of eaters with three kinds of appetites:

✔ Big eaters want two Godzilla hamburgers and three orders of fries.

✔ Regular eaters want one Godzilla hamburger and two orders of fries.

✔ Snackers want no hamburger but one order of fries.

For this example, say you have three big eaters, four regular eaters, and two snackers. Develop a formula that reduces the food order to its simplest terms and then determine how many of each food product you need.

1. **Write a formula for the total order.**

   Write down all the information you have — any or all parts of the problem may come in handy:

   Total order = Big eater orders + Regular eater orders + Snacker orders

2. **Substitute symbols for the words for the eaters.**

   This step makes the math easier. Let T equal the total order, B represent big eaters, R represent regular eaters, and S represent snackers. Rewrite the formula:

   $T = B + R + S$

3. **Write formulas for the contents of each type of order**

   Big eater order = 2 hamburgers and 3 fries

   Regular eater order = 1 hamburger and 2 fries

   Snacker order = 0 hamburgers and 1 fries

4. **Substitute symbols for the food product words and rewrite the formulas.**

   Let h represent hamburgers and f represent fries.

   $B = 2h + 3f$

   $R = 1h + 2f = h + 2f$

   $S = 0h + 1f = 1f = f$

   Note: It's okay to write *zero hamburgers* as *0h*. Multiplying anything by 0 is 0, and this term drops out of the equation. You're also allowed to write *one hamburger* as *1h* and *one fries* as *1f* because multiplying a number or symbol by 1 is that number. But it's better to drop the 1; it's a common convention of algebra not to include the 1.

   The *B*, *R*, and *S* symbols each represent a different combination of hamburgers and fries for the three kinds of eaters.

5. **Rewrite the initial "$T =$" formula by using the "recipes" for the three types of eaters.**

   $$T = B(2h + 3f) + R(h + 2f) + S(f)$$

   This formula is the mathematical representation of how you determine the lunch order. You need only to insert the number of each kind of eater, and you have the results.

6. **Insert the number of each kind of eater and multiply.**

   Use distributivity:

   $$T = 3(2h + 3f) + 4(h + 2f) + 2(f)$$
   $$T = (6h + 9f) + (4h + 8f) + (2f)$$

7. **Use associativity and commutativity to rearrange the items.**

   Group the hamburgers together and group the fries together. Get rid of the parentheses and add up the like quantities — the hs and the fs

   $$T = 6h + 9f + 4h + 8f + 2f$$
   $$T = 6h + 4h + 9f + 8f + 2f$$
   $$T = (6h + 4h) + (9f + 8f + 2f)$$
   $$T = 10h + 19f$$

   The total order *(T)* is 10 hamburgers (10h) and 19 fries (19f). Collect the money and head on out. On the drive, prepare your answer for the most baffling question regarding large fast food orders: "Is that to eat here or take out?"

This story problem isn't trivial. "Burgers and fries" is an analog for any part management or piece management math where you need to find total quantities from individual kits or make up various individual kits from storeroom inventory.

Chapter 14

Quick-and-Easy Geometry: The Compressed Version

In This Chapter

▶ Discovering points, lines, and angles

▶ Looking at simple geometric shapes

▶ Investigating the world-famous Pythagorean theorem

*1*t's useful to know a little geometry, and that's what this chapter contains — a little geometry. *Geometry* is the branch of mathematics that deals with the details of shapes.

Geometry is important because it's conceptual; it improves your thinking. It's part of a full program of math studies, and it has lots of good vocabulary words. And geometry is visual, unlike algebra and trigonometry, so it's easier to grasp. But above all, geometric concepts and words come up in your everyday work.

A flat surface, such as a piece of paper, is called a *plane*, and basic geometry is also called *plane geometry*. You can find other geometries besides plane geometry, such as solid geometry, spherical geometry, Riemannian geometry, Poincaré geometry, and taxicab geometry, which aren't (mercifully) part of this chapter.

In this chapter, you review the basic parts of basic geometry — points, lines, angles, planes, and a coordinate system. And you get the straight skinny about the centerpiece of geometry, the Pythagorean theorem. (Just saying "Pythagorean theorem" makes you feel instantly smarter.)

The many meanings of geometry

Geometry comes from the Greek *geo* meaning "earth" and *metria* meaning "measurement." It started off as earth measurement and then extended to the positions of the planets and the stars. A Greek named Euclid of Alexandria put geometry on the map around 300 BC. He wrote a work called *Elements,* which became the world's bestselling nonreligious work. Euclid set the standard, and to this day geometry is also known as Euclidean geometry.

Today, geometry means different things to different people. Geometry can be

✔ The part of your earlier education where you and math parted ways. Did you leave geometry or did geometry leave you?

✔ A very useful skill in several careers, including carpentry (especially high-end woodworking), graphic design, bricklaying, surveying, and even cosmetology (faces have geometric shapes).

✔ A more sophisticated look at the world than simple shapes and areas (head to Chapter 15 for more on this topic).

✔ An important steppingstone to trigonometry (which we cover in Chapter 16).

Looking at Geometry's Basic Parts

Half of life is knowing what things are called. The other half of life is knowing how things work. Geometry has just a few terms and a few operations. The basic parts (what things are called) are the point, the line, and the angle, plus the plane to draw them on. Finally, you use a coordinate system to describe just where on the plane the parts are located.

Table 14-1 shows where some of these parts fit in dimensional space with other parts:

Table 14-1	Relationships among Different Geometric Elements	
Name	*Dimensions*	*Measurement*
Point	0	Position
Line	1	Length
Plane	2	Area
Cube	3	Volume
Time	4	Volume × time

In Table 14-1, you see volume and time as the third and fourth dimensions. We cover the practical side of these operations in Chapters 15 and 18, so we only deal with them here as they relate to the trades.

A mathematician who works in the field of geometry is called a *geometer*. However, a scientist who works in the field of temperature is *not* called a *thermometer*.

No snakes on this plane: Cartesian coordinates

A *plane* is a flat, two-dimensional surface. It's theoretical, and that means it's perfectly flat and extends forever in all directions. In real life, you draw geometric figures on a flat piece of paper or a flat computer screen.

A *coordinate system* is a way of describing the position of any object on a plane. The most famous and commonly-used coordinate system is Cartesian coordinates. (That's car-*teez*-e-an.) Figure 14-1 shows the Cartesian coordinate system.

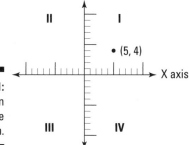

Figure 14-1:
Cartesian
coordinate
system.

The system has two *axes* (that's plural for *axis*). The horizontal axis is the X axis, and the vertical axis is the Y axis. Along each axis are points, and the two axes cross each other at point (0,0). That's the *origin*. As a bonus, you get four *quadrants*, named I, II, III, and IV.

You can describe any position on the plane by naming coordinates. In Figure 14-1, the point is at coordinate (5,4). The two coordinates form an *ordered pair*.

Are you coordinated? You may be in strange territory with a coordinate system, except that coordinate systems are all around you:

- Just about every real estate agent and delivery person has used a *Thomas Guide* at one time or another. This elaborate, multipage map has coordinates on every page. In many ways, the Thomas Guide is better than a GPS.

- Even if you haven't used Thomas maps, you may have used maps from an automobile club. They have coordinates around the edge of the map. The street you want is somewhere in square G-3.

- If you've ever used a GPS for work or for your hobby of geocaching, you've used a coordinate system where positions around the earth are expressed in degrees, minutes, and seconds.

- If you live in an urban center where the streets are on a grid, you're using a coordinate system. If you tell people that you live at 23rd and N in midtown Sacramento, California, you're saying that you're about 19 blocks down and 23 blocks over from the start of the grid.

What's the point?

The *point* is the starting . . . er . . . point for all geometry. A point is an object with zero dimensions. Euclid said a point was "that which has no part." It has no length, height, or thickness. All it has to show for its trouble is a coordinate. For example, (0,0), (5,3) and (29.5645,56.1) are all points in a coordinate system.

A point is the smallest and most precise entity in geometry. It's pivotal to using all the other entities: the ray, the line segment, the angle, the curve, and the geometric shape. The point always represents the smallest, sharpest, or most fundamental unit in your activities. Here is a partial list of points you use in work and life:

- Smallest unit of measurement for a gemstone
- Smallest unit of measurement for the thickness of paper
- A part of a home loan
- A separator in decimal numbers
- The sharp end of a pencil
- The lead soldier in a combat patrol
- A projection of land into the ocean
- A direction on a compass
- A unit in scoring games

What's your line?

A *line* is a geometric element with 1 dimension, length. It has no thickness and no height. The line is a big advance over the point, although the line denies using steroids.

In geometry, a line is conceptual, not real, so it's infinitely long and perfectly straight. Figure 14-2 shows a line and two variations.

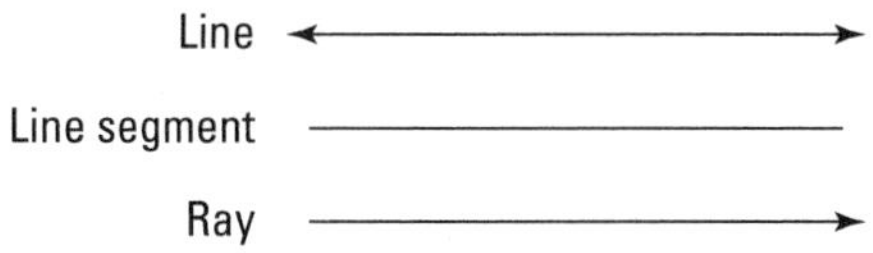

Figure 14-2:
A line
and two
variations.

Line

Line segment

Ray

The first element in Figure 14-2 is a line. It's infinitely long, and that's why it has two arrowheads. If this book were infinitely wide, you'd see the whole line.

The second element in Figure 14-2 is a *line segment,* which is a portion of a line. Each end of a line segment has a coordinate, such as (0,0) and (3,4). Line segments make up squares, rectangles, triangles, and other geometric shapes (which we cover later in the chapter).

The third element in Figure 14-2 is a *ray.* It's a mix of line segment and line. It has a starting point, but the other end goes to infinity. You use rays to describe angles formally.

Parallel lines are two lines that go on side by side forever and never meet. *Perpendicular* lines are two lines that meet at a 90-degree angle, which we just so happen to discuss in the following section.

What's your angle?: Acute, obtuse, and right angles

Put on your tuxedo and get ready for something formal. In geometry, formally, an *angle* is a figure with two rays joined at the same endpoint. (Check out the preceding section for more on rays.) The endpoint is called the *vertex.* The size or amount of the angle is called the *magnitude.* Figure 14-3 illustrates an angle.

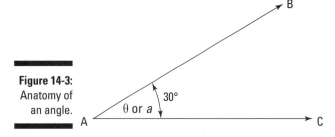

Figure 14-3:
Anatomy of
an angle.

The angle in the illustration is called *a*. Sometimes textbooks give the angles as Greek letters. In this example, it's theta (θ).

You also call the angle in the illustration BAC, because BAC summarizes the points that make it up. A is the vertex. B and C are points on the rays. You write this name as ∠BAC or ∠CAB. The vertex point is always in the middle.

The magnitude of the angle in the illustration is 30 *degrees* (30°). The magnitude is the portion of a circle (360 degrees) that the angle sweeps through. There's another system of measurement, called radians, but you aren't likely to encounter radian measurement.

If you're a carpenter, you use angles when you determine roof pitch (also called slope or angle), but you express the angle differently. If you frame a "7/12" roof, the roof rises 7 inches for every 12 inches it runs. You may also call it a *pitch* of "7-12," "7 to 12," "7 and 12," or "7 on 12." No matter what you call it, the pitch has an angle of 30.5 degrees. Similarly, surveyors and heavy equipment operators (road builders) deal with angles in road grade inclines and declines, but their terminology includes the word *grade*, not angle. A 5-percent grade rises 5 feet for each 100 feet of travel, or about 264 feet in a mile. A 5-percent grade has an angle of 2.9 degrees.

You can calculate roof pitch and road grade using trigonometry (see Chapter 16), but you get it faster using Internet calculators.

Angles have special names. The three common angles you encounter:

- ✔ *Acute angles* (less than 90 degrees) are sharp angles.

- ✔ *Right angles* (exactly 90 degrees) are square corners. They're marked by a little box in the angle, as you can see in Figure 14-4.

- ✔ *Obtuse angles* (more than 90 degrees but less than 180 degrees) are dull angles.

Figure 14-4 shows several types of angles.

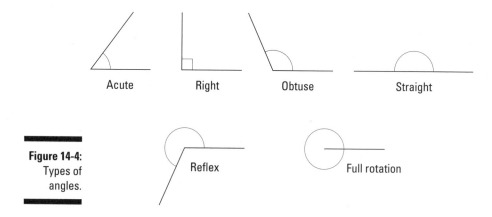

Figure 14-4:
Types of
angles.

The examples of straight, reflex, and full rotation angles in Figure 14-4 illustrate other angle possibilities. However, acute, right, and obtuse angles make up 99.999 percent of the work you do in real life and math classes. Many other angle types exist, but they're meaningful mainly to mathematicians.

If you've got to know, the other angle types include oblique angles, congruent angles, vertical angles, adjacent angles, complementary angles, supplementary angles, explementary (or conjugate) angles, interior angles and exterior angles.

Examining Simple Geometric Shapes

A simple geometric shape is a two-dimensional shape that consist of points, lines, curves, and planes. There are many geometric shapes (hundreds, in fact), but the shapes you use in everyday work (and we cover in the following sections) are the square, rectangle, triangle, and circle.

The square and the rectangle

Speaking formally, a *square* is a rectangular quadrilateral geometric shape. It has four equal sides and four equal angles, which are always right angles. A square is a special kind of rectangle.

So what's a *rectangle?* It has two sets of sides (two equal sides for length and two equal sides for height), but unlike the square, its length and width aren't always equal to each other. The angles are right angles.

Figure 14-5 shows a square and a rectangle.

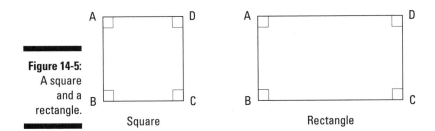

Square

Rectangle

The small boxes in the corners indicate right angles. If you write labels A, B, C, and D at the corners, you then refer to the figures as ABCD.

The triangle: Just because it isn't a right triangle doesn't mean it's wrong

A *triangle* is a simple geometric shape with three sides (well, line segments). A triangle contains three angles, which certainly explains the name *triangle*.

Since ancient times, triangles (along with squares and rectangles) have been used to lay out fields for farming — and to measure them for taxes. (You can find more on measuring the area of a triangle in Chapter 15.) Like angles, triangles have special names. In fact some of the triangle names are identical to angle names. Figure 14-6 shows several types of triangles.

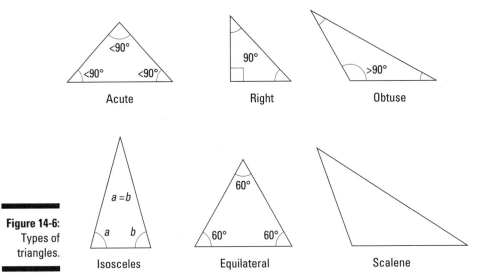

Acute

Right

Obtuse

Isosceles

Equilateral

Scalene

The examples in the figure show every condition you encounter with triangles. If a triangle doesn't behave like one of those samples, it's not a triangle.

- ✔ *Acute triangles* have three acute angles.
- ✔ *Right triangles* have one 90-degree angle.
- ✔ *Obtuse triangles* have one angle that's more than 90 degrees.
- ✔ *Isosceles triangles* have two equal angles.
- ✔ *Equilateral triangles* have three equal (60-degree) angles.
- ✔ *Scalene triangles* have three different angles.

The first triangle in Figure 14-6 is labeled ABC. You express the triangle's name with a small triangle sign (that is, △ABC) and the letters ABC.

A right triangle is part of the Pythagorean theorem, and it has special names for its parts. The two sides next to the right angle are the *legs* and the diagonal is called the *hypotenuse*. Check out "Learn It Once and Forget It: The Pythagorean Theorem" later in the chapter for more on this formula.

The polygon

A *polygon* isn't a parrot who flew the coop. It's a figure with many sides. A *regular* polygon has equal angles and equal sides. An *irregular* polygon has sides of different lengths and angles with different values. You can find other types of polygons, too, including concave and star-shaped. Figure 14-7 illustrates several regular polygons and one irregular polygon.

Figure 14-7:
Regular and
irregular
polygons.

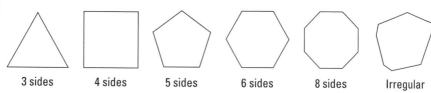

| 3 sides | 4 sides | 5 sides | 6 sides | 8 sides | Irregular |

You use a few polygons so much that they have common names. For example:

- ✔ A five-sided regular polygon is a *pentagon*. That's also the name of a very large five-sided office building in Washington, D.C.
- ✔ A six-sided polygon is a *hexagon*. It's also the shape of each cell in a bee's honeycomb.
- ✔ An eight-sided polygon is an *octagon*.

Now when you drive down the road, you can point to the signs and shout, "That stop sign is an octagon!" Unfortunately, this valuable piece of math information may not make you any more popular with friends and may disturb strangers.

The circle

A *circle* is a curved geometric shape where every point on the curve is an equal distance from a fixed point called the *center*. Circles are everywhere: the sun, the moon, bowls, plates, tires and steering wheels, just to name a few. Figure 14-8 shows the parts of a circle.

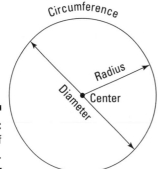

Figure 14-8:
Anatomy of
a circle.

The circle has the following parts:

- ✔ The *circumference* is the length of the line segment that constitutes the circle. It's the distance around the circle.

- ✔ The *diameter* is the distance across the circle through the center.

- ✔ The *radius* is the distance from the center to the edge of the circle.

The great thing about a circle is that when you know the radius you know the diameter (it's twice as much), and when you know the diameter you know the radius (it's half as much).

When you multiply the diameter by pi (π) you know the circumference. When you divide the circumference by pi, you know the diameter.

When you know the radius, square it and multiply by pi, you know the area of the circle. Chapter 15 has the details on finding area.

By convention, a circle has 360 degrees (the same unit of measure you use for angles). A 90-degree angle sweeps one quarter of a circle, a 180-degree angle sweeps half, and a 360-degree angle (full rotation in Figure 14-4 earlier in the chapter) is a full circle. Flip to "What's your angle?: Acute, obtuse, and right angles" earlier in the chapter for more on angles.

Learn It Once and Forget It: The Pythagorean Theorem

The Pythagorean (that's puh-thag-o-*re*-an) theorem is important to know, mostly for educational reasons. It shows you a bit of geometry that's supported by algebraic analysis, and it's the first of several more sophisticated algebra, geometry, and trigonometry principles you may need as part of your professional education. It's part of the pathway to greater understanding of and comfort with math. Later, you may use the theorem occasionally in a surveying-related career, but for many other careers, you may not need it again. It's important to understand, but not important to remember.

The *Pythagorean theorem* states that the sum of the squares of a right triangle's leg lengths is equal to the square of the length of its hypotenuse. Figure 14-9 shows a classic view of the theorem.

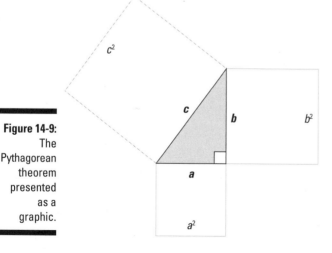

Figure 14-9:
The Pythagorean theorem presented as a graphic.

In Figure 14-9, the a^2 represents the square of length a. And b^2 is the square of length b. Also, c^2 is the square of side c, the hypotenuse.

In algebra, you write

$$a^2 + b^2 = c^2$$

And (drum roll, please) after you add a² and b² to get c², you then calculate the square root of c² to find the length of the hypotenuse.

Thanks to the various basics of math, the following two algebra equations are also true:

$$a^2 = c^2 - b^2$$
$$b^2 = c^2 - a^2$$

Bottom line: If you know the lengths of two sides of a right triangle, you can easily calculate the length of the third side.

Try this concept for yourself. In Figure 14-10, one leg has a length of 2 and the other leg has a length of 4. Calculate the length of the hypotenuse by using the Pythagorean theorem.

Figure 14-10:
Use the
Pythagorean
theorem
to find the
missing
length.

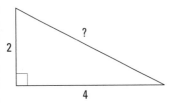

Start with the formula:

$$a^2 + b^2 = c^2$$

Substitute the values you know:

$$2^2 + 4^2 = c^2 \text{ or } 4 + 16 = c^2$$
$$20 = c^2$$

Get the square root of 20 from a calculator, smartphone, or spreadsheet program. The answer is 4.4721.

Play it again, Samos

Pythagoras (c. 570–c. 495 BC), the creator of the Pythagorean theorem, was born on the island of Samos. But he's not its only famous native son. The Greek philosopher Epicurus was born on Samos in 341 BC, and famous fabler Aesop (620–560 BC) may also have been born on Samos, too. He certainly lived there.

The Pythagorean theorem has been proved formally many times by skilled mathematicians in all the ages. What people don't know is that as a young man U.S. President James A. Garfield did a clever and unusual proof of the Pythagorean theorem, called the trapezoidal proof. Before he got into politics, Garfield was an outstanding student.

The theorem even has its place in classic film. Take a look at this great quotation from "The Wizard of Oz" (1939).

> Scarecrow: The sum of the square roots of any two sides of an isosceles triangle is equal to the square root of the remaining side. Oh joy! Rapture! I got a brain! How can I ever thank you enough?
>
> Wizard: You can't.

Of course, the Scarecrow had the Pythagorean theorem completely wrong, but don't let that ruin your enjoyment of the movie.

Example: Don't Fence Me In

You work for a fencing contractor. As a result, you are part artist, part business person, part post hole digger, and part mathematician. Your client has a stable in horse country and wants you to install vinyl horse fencing for its pastures. You need to calculate the fencing requirements for the ranch.

Figure 14-11 shows the pasture areas.

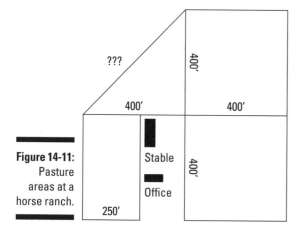

Figure 14-11: Pasture areas at a horse ranch.

You have most of the dimensions for the four pastures, and they're easy enough to add up. But what about that diagonal (marked ??? in the figure)? How long is it? Here's how you find out:

1. **Recognize that the pasture in question is in the shape of a right triangle and that you want to know the length of its hypotenuse.**

   Your Pythagorean theorem alarm should be going off — that's the formula you need here.

2. **Determine the values you need to use and insert them into the formula.**

   One leg of the triangle is 400 feet. The other leg of the triangle is also 400 feet, so input those into the formula.

   $$c^2 = 400^2 + 400^2$$

3. **Calculate the value of c^2.**

   $$c^2 = 160{,}000 + 160{,}000 = 320{,}000.$$

4. **Use a calculator to find the square root of c^2.**

   $$c = 565.7 \text{ feet}$$

   The fencing run for the diagonal part of the pasture is 565.7 feet.

Example: The Pen Is Mightier Than the Paddock

The client from the preceding section also wants a three-rail vinyl fence for a 50-foot, circular exercise pen. How much rail should you get to do the job?

Figure 14-12 shows the exercise pen.

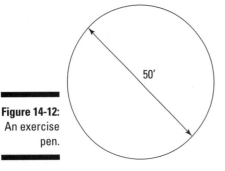

Figure 14-12:
An exercise pen.

50′

1. **Recognize that you're installing fence for a round plot and therefore need to find the circumference.**

   You know you can use the formula $c = \pi d$ when you know the pen's diameter.

2. **Insert the diameter and the value of pi in the formula and calculate the value of *c*.**

   Use 3.14159 for pi.

   $$c = 3.14159 \times 50$$

   $$c = 157 \text{ feet}$$

   That's the circumference. It takes 157 feet of rail to put one rail around the pen.

3. **Multiply the circumference by three to account for the three rails of the fence.**

   This fencing run calls for three rails; use the formula.

   $$\text{Total feet of rail} = 157 \text{ feet/rail} \times 3 \text{ rails}$$

   You need 471 feet of rail.

Chapter 15

Calculating Areas, Perimeters, and Volumes

In This Chapter

▶ Identifying shapes and calculating their areas by using easy formulas

▶ Knowing the parameters of perimeters (and figuring out their lengths)

▶ Performing volume calculations

A lot of math (technical and otherwise) is about calculating areas, perimeters, and volumes. Those calculations can be fun, easy, and very useful. For example, computing areas is right at the core of the building trades. If you do carpentry, concrete, painting, carpeting, or drywall, you use area calculations a lot. You work with perimeters in fashion technology, landscaping, and fence installation, among other fields.

The lab tech, the cosmetologist, the brewer, and the chef deal with volumes more than other careers do, but they aren't the only ones. Automotive technicians have to know how to figure engine capacity, and cement masons need to know concrete volumes.

There are two wonderful things, in particular, about areas, perimeters, and volumes. The first is that the handiest, most important calculations are the easiest ones. The second is that you grew up with some of the basic shapes they deal with, so you've got a head start.

In this chapter, you review the entire length and width of areas, work your way around perimeters, and explore the not-so-voluminous world of volumes.

Area: All That Space in the Middle

An *area* is a region surrounded by a *closed curve* (an entity with a continuous border, like a square, a rectangle, a triangle, or a circle). Figure 15-1 shows closed and open curves.

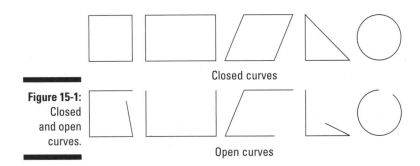

Closed curves

Figure 15-1:
Closed
and open
curves.

Open curves

Even if the shape you're looking at is all made up of straight lines, it's still technically a closed curve. And that means you can calculate its area.

In mathematics classes, an area is usually just a number, but in real life, you have to attach units of measurement. For example, you express carpeting and floor areas in square feet. A printer expresses paper density in grams per square meter (g/m^2). If you're surveying or mapping, you use square miles, sections, and townships. Flip to Chapter 6 for other units of square measure.

According to the U.S. Department of the Interior, the Public Land Survey System started soon after the Revolutionary War. Over the past two centuries, almost 1.5 billion acres in the United States have been surveyed into townships and sections. A *section* is one square mile and contains 640 *acres*. Thirty-six sections make up a *township* of 36 square miles.

When you do real-life area calculations, make sure that every unit of measurement is the same. If you're calculating the area of a square and encounter one side measurement in feet and the other side measurement in yards, you need to convert one of the units before you start calculating. Chapter 6 shows you how.

Calculating the area of rectangles and squares

A *rectangle* is a region with four sides and four 90-degree angles. The two length sides (long sides) have equal measurements, and the two width sides (short sides) are equal to each other as well. Figure 15-2 shows a rectangle.

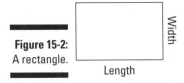

Figure 15-2:
A rectangle.

Width

Length

The area formula for a rectangle is length multiplied by width. You say this statement as "Area equals length times width." You write it as

Area = Length × Width

$A = L \times W$

You encounter many rectangles when you calculate paint for interior walls. For example, a walk-in closet wall may be 6 feet wide and 8 feet tall. Figure 15-3 shows the wall.

Figure 15-3: A wall in a walk-in closet.

6 feet / 8 feet

In Figure 15-3, we play fast and loose with terms like *length, width,* and *height.* As long as one of your dimensions is the *L* dimension required by the area formula and one is the *W* dimension, the calculation comes out the same.

Find the area of Figure 15-3 by plugging those measurements into the formula for the area of a rectangle.

$A = 6 \text{ feet} \times 8 \text{ feet}$

$A = 48 \text{ square feet}$

Of course, any interior painting job likely involves several walls in a room and several rooms in a house, plus ceilings. Having a handle on the basic area calculation allows you to calculate the full area you're painting to help you determine how much paint you need.

The good news (if you're a painter, anyway) is that you can use shortcuts. A professional house painter may calculate area based on the *total* lengths of all the walls in a house and assume a height of eight feet.

Painting isn't the only real-life situation where you may encounter rectangle-area calculations. If you go into business for yourself and rent office space, you rent by the square foot. If you buy a house, the listing and the appraisal reflect square feet. Square footage is also important when you calculate (for income tax purposes) the portion of your home used for office space.

A *square* is a special rectangle that has four equal sides; it's the only rectangle where the length and width are the same. Figure 15-4 shows a square.

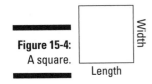

Figure 15-4:
A square.

Width

Length

To figure the area of a square, you recycle the formula for the area of a rectangle: $A = L \times W$.

Figure 15-5 applies to a small room that needs carpeting or flooring, or a tiny patio that needs a concrete slab.

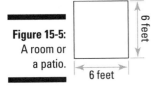

Figure 15-5:
A room or
a patio.

6 feet

6 feet

You have the formula and you have the dimensions. You can now use the time-honored mathematical method of "plug and chug" to get the answer.

A = 6 feet $\times$ 6 feet

A = 36 square feet

Figuring the area of a parallelogram (a bent-over long rectangle)

A *parallelogram* is a region with four sides. The two length sides are equal to each other and the other two sides are equal to each other, which makes the opposite sides parallel to each other.

If you're asking yourself why a parallelogram isn't a rectangle (see the preceding section), remember that a rectangle requires four right angles. A parallelogram doesn't have 90-degree angles — it looks like a squooshed rectangle. Figure 15-6 shows a parallelogram.

Figure 15-6:
A parallelo-
gram.

Height

Base

The formula for the area of a parallelogram is a slight variation of the formulas for squares and rectangles: You multiply the base *(B)* by the height *(H)*. You say this calculation as "Area equals base times height" and write it as

Area = Base × Height

$A = B \times H$

Calculating with the length of the short side won't get you there. The *height* (the distance from the base to its opposite side) is what matters.

You don't encounter a surface that's an exact parallelogram every day, but some patios, driveways, and walkways come pretty darned close. If three sides match those of a parallelogram, you can (hypothetically) make the fourth side into the fourth "correct" side as shown in Figure 15-7. In technical terms, this tactic is called "cheating" or "pretending."

Figure 15-7 shows the shape of a new patio to be poured. It abuts an existing circular swimming pool and is approximately the shape of a parallelogram.

Figure 15-7:
A patio
in the shape
of a paral-
lelogram.

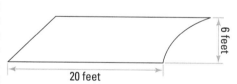

The parallelogram isn't perfect, but it's close enough to what you need to allow you to fudge to get an area. Using the measurements from Figure 15-7 and the formula for the area of a parallelogram, you now find the area.

$A = 20$ feet × 6 feet

$A = 120$ square feet

Determining the area of a trapezoid (a trapewhat?)

A trapezoid, that's what. A *trapezoid* is a region with four sides, but it has only one set of parallel sides (called the *bases*). Figure 15-8 shows a trapezoid.

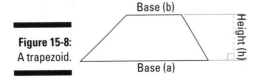

Figure 15-8:
A trapezoid.

Finding the area of a trapezoid takes a little more information than finding the area of a rectangle or a parallelogram does. (We cover those shapes in the preceding sections.) You need to know the length of *both* of the bases (labeled *a* and *b* in Figure 15-8) and the height (labeled *h* in Figure 15-8).

Like in a parallelogram, the trapezoid's height is important; however, you calculate the *average* length of the two bases for a trapezoid.

In a trapezoid, area equals height times half the sum of Base a and Base b. And that's the way you say it, too.

$$\text{area} = \text{height} \times \frac{\left(\text{base } a + \text{base } b\right)}{2}$$

$$A = h \times \frac{\left(a+b\right)}{2}$$

Figure 15-9 shows a trapezoid with its dimensions marked.

Figure 15-9:
A trapezoid with dimensions.

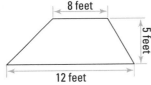

Because you know the information for this trapezoid, plug that information into the formula to find the area.

$$A = h \times \frac{\left(a+b\right)}{2}$$

$$A = 5 \times \frac{\left(8+12\right)}{2}$$

$$A = 5 \times \frac{\left(20\right)}{2} = 5 \times 10$$

$$A = 50 \text{ square feet}$$

$$A = 50 \text{ ft}^2$$

A quadrilateral family tree

Rectangles, squares, parallelograms, and trapezoids are all *quadrilaterals*, which means they're four-sided polygons. Here is the genealogy of the family headed by the quadrilateral: A trapezoid is a special offshoot of a quadrilateral — two sides are parallel. A parallelogram is a special offshoot of a trapezoid — it has equal, parallel opposite sides. A rectangle is a special offshoot of a parallelogram — it has right angles. A square is a special offshoot of a rectangle — it has right angles and equal sides.

Calculating the area of a triangle

A *triangle* is a three-sided figure with three angles, such as the one shown in Figure 15-10. Math people used to say that a triangle was just half of a rectangle, but that's not always the case. If you have an *obtuse* triangle (a triangle where one angle is more than 90 degrees) or an *acute* triangle (where all angles are less than 90 degrees), the triangle is half a parallelogram, not a rectangle. Only a right triangle (a triangle with one angle of 90 degrees) is half a rectangle. (Check out Chapter 14 for more on the various kinds of triangles.)

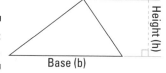

Figure 15-10:
A triangle.

Height (h)

Base (b)

Figuring the areas of triangles can come up in real estate or surveying when you determine lot sizes. The size of the lot can be very important in towns with strong zoning commissions because the property area devoted to parking is often a critical issue.

Just like your favorite recipes and a few home cures for the cold, the formula for figuring out a triangle's area is simple: Area equals one-half the product of the base and the height. You write this formula as

$$\text{Area} = \frac{1}{2} \times \text{base} \times \text{height}$$

$$A = \frac{1}{2} \times b \times h \quad \text{or} \quad A = \frac{1}{2}bh$$

Figure 15-11 shows you a sample triangle with its dimensions labeled.

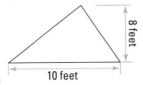

Figure 15-11:
A triangle with marked dimensions.

8 feet

10 feet

You know the formula and the dimensions, so just calculate.

$$A = \frac{1}{2} \times b \times h$$

$$A = \frac{1}{2} \times 10 \text{ ft} \times 8 \text{ ft}$$

$$A = 40 \text{ ft}^2$$

TECHNICAL STUFF

The mathematical mind loves names, and triangles are no exception; the list of triangle names rivals the flavors in a small ice cream parlor: equilateral triangles, isosceles triangles, scalene triangles, right triangles, acute triangles, and obtuse triangles.

Computing the area of a circle

A *circle* is a curved region with an equal distance from a fixed point (the *center*) to any point on the edge. A circle is the simplest area you'll ever do. You need only one measurement: the radius. The *radius* is the distance from the center of the circle to any point on the curved edge of the circle; it's the distance halfway across a circle. The full distance across (through the center) is called the *diameter*. It just happens to be twice the length of the radius, so when you know the diameter, you can easily get the radius. Figure 15-12 illustrates a circle with its radius.

Figure 15-12:
A circle with its radius.

Radius

In addition to the radius, you also need *pi* (a Greek letter pronounced "pie" and represented by the symbol π) to calculate a circle's area. Pi is called a *mathematical constant* because its value doesn't change.

It's an oval! It's an egg! No, it's an ellipse

An *ellipse* is like a circle, but it's not. Like a circle, it's a region surrounded by a closed curve, but it's taller or shorter than it is wide. It's a squooshed circle. The formula for calculating the area of an ellipse is similar to that for a circle.

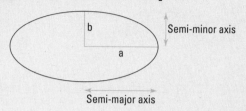

$A = \pi \times a \times b$ or $A = \pi ab$

In the formula, *a* is the semi-major axis and *b* is the semi-minor axis, which you can see in the following figure. The *semi-major axis* is one half of the *major axis* (the total length of the ellipse). The *semi-minor axis* is one half of the *minor axis* (the total height of the ellipse).

A famous ellipse in the U.S. is a giant lawn or green in President's Park in Washington, D.C. It's about 17 acres in area, with a length of 1,058.26 feet and a width of 902.85 feet. That may make it the largest ellipse in the world.

Your calculator may have a π key. If it doesn't, use 3.14159, which is an approximation.

The formula for finding a circle's area is simple: Area equals pi times the radius squared. You write this calculation as

Area = π × radius squared

$A = \pi \times r^2$ or $A = \pi \times r \times r$ or $A = \pi r^2$

Pi is an *irrational number,* which means it goes on forever and can't be concretely written. You can't express it as ratio. Supercomputers have worked out π to over 1 trillion digits, and that's still not the end.

And now for a sample calculation. Drumroll, please! Figure 15-13 is a circle with one dimension, the radius.

Figure 15-13: A circle with a radius of four feet.

$r = 4$ feet

Using Figure 15-13 and the area formula, you can now calculate the area.

$A = \pi \times 4^2$

$A = \pi \times 16$

$A = 3.14159 \times 16$

$A = 50.27$ square feet

In the construction trades and in fashion design, you make measurements of circles and parts of circles from time to time. The example in Figure 15-13 is a little small for a patio and a little large for a tablecloth. But the idea is that you get the idea.

Perimeters: Along the Edges

Areas (see the preceding sections) aren't the only shape measurements you may need in your work. In a number of careers, you need to know "how far around" an area is, which is a clumsy way of saying that you need to know how to calculate perimeters.

Figuring out perimeters is simple. The units may vary (for example, you use feet to measure the perimeter of a fencing job and miles to measure the perimeter of a state), but the principles are identical. Figuring perimeters isn't very glamorous, and most people don't think it's exciting. It's like kissing: When you're young, at first it doesn't seem like much fun, but later it's more appealing. Can't say the same for broccoli or Brussels sprouts. So if you're ever given a choice between calculating perimeters and eating those vegetables, take the perimeters. The following sections give you the lowdown on finding the perimeters of basic shapes, including the square, rectangle, parallelogram, trapezoid, triangle, and circle.

Understanding perimeters: What goes around comes around

To be official about it, the *perimeter* is the outer limit of a region that's bounded by line segments. Good, because every shape we discuss here is a region bounded by line segments. To put it another way, the perimeter is the path around an area. The word comes from the Greek, where *peri* means "around" and *meter* means "measure."

Figure 15-14 illustrates a perimeter. In this case, the perimeter is the distance around all four sides of a rectangle.

Figure 15-14:
The parts
of a rect-
angle's
perimeter.

If you're landscaping, think bender board. If you're pouring concrete, think forms. If you're designing clothing, think rick rack. Even physical therapists and fitness trainers measure around waistlines, biceps, and thighs. Those are perimeters, too.

Calculating the perimeters of polygons

The bad news is that we can't give you a universal "plug and chug" formula for all polygon perimeters. The good news is that we can give you several easy formulas for them.

The closest you can find to a general formula is the following:

$$P = s + s + s + s + s + s + \ldots$$

All this formula is saying is that the perimeter *(P)* equals the sum of all the sides *(s)*, just as you see in Figure 15-14. The ellipsis (. . .) in the formula means you keep on adding until you've got all the sides accounted for.

But instead of all that adding, follow these shortcut formulas:

- ✔ **Square:** Multiply one side by four.

- ✔ **Rectangle:** Multiply the length by two, multiply the width by two, and then add the products.

- ✔ **Parallelogram:** Multiply the length (long side) by two, multiply the short side by two, and then add the products.

- ✔ **Trapezoid:** Add all four sides.

- ✔ **Triangle:** Add all three sides.

- ✔ **Regular polygon:** For regular polygons (polygons with equal sides), multiply the length of one side by the total number of sides. For example, multiply one side of a regular pentagon by five to determine the shape's perimeter.

- ✔ **Irregular polygon:** Add all the sides.

A perimeter by any other name: Finding a circle's circumference

Circles are kind of like the black sheep of the perimeter family. A circle doesn't have a perimeter — it has a circumference. The *circumference*, just like a perimeter, is the distance around the curved edge of a circle. The word comes from the Latin *circumferentia*, from *circum* ("around") and *ferre* ("to carry").

Use this great formula to find the circumference of a circle.

Circumference equals pi multiplied by the diameter

Circumference = $\pi \times$ diameter

$C = \pi \times d$ or $C = \pi\, d$

The variable d is the diameter of the circle, and π is pi (about 3.14159).

Figure 15-15 is an example of finding a circumference.

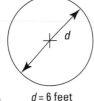

Figure 15-15:
A circle and
its diameter.

$d = 6$ feet

Plug the information from Figure 15-15 into the formula to find the circumference.

$C = \pi \times d$

$C = \pi \times 6$

$C = 18.85$ feet

Volume: The Third Dimension

You can't turn the knob down on these volumes, the volumes of rectangular solids and spheres. But you can find the *volume,* the amount of three-dimensional space a shape takes up.

You express volume in cubic units (for example, cubic feet). In fact, volume is so important to so many vocations (especially science) that many fields have specialized units of volume. Cooking has many traditional volume measurements, while the laboratory uses metric volume units. And the most common volumetric unit is in front of you every time your refuel your car. It's called a gallon. Head to Chapter 6 for more on units of volume.

Getting a handle on American volume units

Volume measurement, conversion, and calculation are often essential to your professional life, so you need to be familiar with the terminology you encounter. The two main systems of measurement (for measuring volume and otherwise) are the American system and the metric system, which we cover in Chapter 6. American volume measurements get special treatment here because they're a little more complex than the relatively straightforward metric volumes and require some explaining.

A standard U.S customary unit for measuring volume is the cubic inch (cu in or in^3). A *cubic inch* is the volume of a cube with one inch on each side. Similarly, a *cubic foot* (cu ft or ft^3) is the volume of a cube that has sides of one foot. A *cubic yard* (cu yd, CY, or yd^3) is the volume of a cube that has sides of one yard. These units are used in only a couple of the trades.

Because a foot contains 12 inches, a cubic foot contains $12 \times 12 \times 12$ cubic inches, or 1,728 in^3.

The cubic inch used to be the standard unit for describing automobile engine capacity (sometimes called *displacement*). It's been overtaken by metric units (the liter and the cc), but it's still used when talking about American muscle cars. When The Beach Boys sang "409" in 1962, they were singing to the dream of American boys — owning a Chevrolet 409, a W-series V-8 engine. It had a displacement of 409 cubic inches.

The U.S. liquid gallon (fl gal) is the main volume unit for liquids in the United States. You often buy apple juice, drinking water, and gasoline in gallons (or some subdivision of a gallon, like a quart or a pint). A cubic foot contains 7.48051948 gallons; by law, the U.S. gallon is equal to 231 cubic inches.

Calculating the volume of cuboids (also known as boxes)

The most-used volume calculation is for the *cuboid,* a solid figure bounded by six faces. That is, it's a box. All its angles are right angles, and its opposite faces are equal. (A *cube* is a special case where all the faces are square.)

Figure 15-16 shows a cuboid.

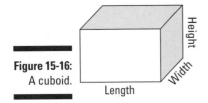

Figure 15-16:
A cuboid.

The formula for calculating the volume of a cuboid is volume equals length times width times height. You write this equation as

Volume = Length × Width × Height

$$V = L \times W \times H$$

Although this formula is very straightforward, you live in an age of electronic automatic wonders and can find box volume calculators on the Internet.

If you're a chef, you likely need to know the capacities of your reach-in refrigerators, usually measured in cubic feet. (Don't rely on the manufacturer's published capacity; manufacturers often count unusable space such as door water dispensers, hardware, and shelves in their calculations.) Figure 15-17 illustrates an example of a reach-in refrigerator.

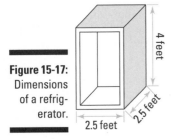

Figure 15-17:
Dimensions of a refrigerator.

You know the dimensions of the interior space. Plug them into your formula to get the volume of the refrigerator.

$$V = L \times W \times H$$
$$V = 2.5 \times 2.5 \times 4$$
$$V = 25 \text{ cubic feet}$$

Finding the volumes of spheres and cylinders

A *sphere* is a round object; balls, globes, and the sun are all common spheres. As time passed and people stopped believing that the earth was flat, the earth became a well-known sphere as well. Figure 15-18 shows a sphere.

Figure 15-18: A sphere.

radius

The formula for figuring a sphere's volume isn't like the formula for a cuboid but rather more like the formula for the area of a circle (which you can find in "Computing the area of a circle" earlier in the chapter). The formula for calculating the volume of a sphere is volume equals four pi times the radius cubed divided by three; you write this calculation as

Volume equals $\frac{4}{3}$ pi times the radius cubed

$$V = \frac{4\pi r^3}{3}$$

Note: Calculating the volume of a sphere isn't super common in the trades, but you never know when it will pop up.

A related shape that comes up in automotive technology is the cylinder.

A *cylinder* is a curvilinear geometric shape. In plain talk, it looks like a can or a box of oats. Like the sphere, the cylinder is circle-based. See the cylinder in Figure 15-19.

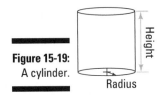

As you may have predicted, there's a formula for calculating the volume of a cylinder. Its volume equals pi times the radius squared times the height, which you write as

Volume = $\pi \times$ radius squared $\times$ height

$V = \pi \times r^2 \times h$ or $V = \pi \times r \times r \times h$ or $V = \pi\, r^2 h$

Example: Bore and Stroke for the Auto Guy

You're a well-trained auto mechanic. You have an excellent set of costly tools and are able to work with automobile diagnostic computers. But more importantly, you love cars. That's why you're ecstatic when one of your clients brings in an old Chevrolet Impala, a classic from the early 1960s, that he just bought. You identify the engine as a Chevy short block V-8, but your client wants to know its capacity so he can determine whether it's a *screamer* or just a so-so engine.

To determine engine capacity, you measure the *bore* (diameter) and *stroke* (height) of the cylinders. This formula is a simple variation of the formula for calculating the volume of a cylinder. The differences are

- ✔ You can measure diameter (bore) but not the radius of a cylinder.

- ✔ You have eight cylinders. Fortunately (being an engine) all the cylinders are the same size.

Use this formula to calculate capacity:

$$\text{Capacity} = \pi \times \left(\frac{\text{bore}}{2}\right)^2 \times \text{stroke} \times \text{number of cylinders}$$

The first part,

$$\pi \times \left(\frac{\text{bore}}{2} \right)^2$$

is just the area of a circle, where you divide the diameter (the bore) by 2 to get the radius and then square the radius and multiply by pi.

The second part, × stroke, asks you to multiply by the height of the engine cylinder, much like the formula for calculating the volume of a cylinder does. The last part, × number of cylinders, is very straightforward.

You measure a cylinder in the Chevy and find that it has the following dimensions:

- ✔ Bore = 4.312 inches
- ✔ Stroke = 3.50 inches

Plug the information you have into the formula and calculate the capacity.

$$\text{Capacity} = 3.14159 \times \left(\frac{4.312}{2} \right)^2 \times 3.50 \times 8$$

$$\text{Capacity} = 3.14159 \times 4.6483 \times 3.50 \times 8$$

$$\text{Capacity} = 408.8855$$

The answer is 409 cubic inches. Your client's car came with one of the classic high-capacity engines. They're going to love him at the classic car show!

Although this formula isn't hard to apply, you can also get the answer through an engine capacity calculator on the Internet.

Example: Yard Area, the Landscaper's Nightmare

You're a landscape contractor. The client wants a new lawn, and you need to calculate its area to estimate for seed or sod. The trouble is, the lawn area isn't your regular garden-variety (har har) area. It has an irregular shape.

Figure 15-20 shows the lawn area and its dimensions.

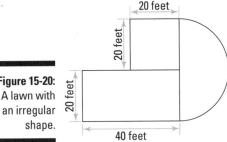

Figure 15-20: A lawn with an irregular shape.

This area is made up of shapes you know. Use the area formulas in this chapter and the following steps to calculate the total area.

1. **Inspect the shape.**

   It's made up of a square, a rectangle, and part of a circle. It looks like half a circle, so you calculate the area of a full circle and divide by two.

2. **Create a formula.**

   You're basically just adding the areas of the component shapes, so your formula looks something like the following:

   Total area = Area of square + area of rectangle + $\frac{1}{2}$ area of circle

3. **Determine the dimensions for the shapes.**

   - The dimensions for the square are given (20 feet × 20 feet).

   - The dimensions for the rectangle are also given (20 feet × 40 feet).

   - The dimension for the circle (the radius) is easy to see on inspection. The radius is the height of the square or the rectangle (20 feet).

4. **Calculate the individual areas.**

   - Area of the square = 20 feet × 20 feet = 400 square feet

   - Area of the rectangle = 20 feet × 40 feet = 800 square feet

   - Area of the circle = $1/2 \times \pi \times 20^2$ feet = 628.3 square feet

5. **Add the areas together.**

   - Total area = 400 square feet + 800 square feet + 628.3 square feet

   - Total area = 1,828.3 square feet

The answer is 1,828.3 square feet.

Chapter 16

Trigonometry, the "Mystery Math"

In This Chapter

▶ Getting a handle on trigonometry basics

▶ Introducing the six common trigonometric functions

▶ Using trigonometry calculations to get real-world results

Trigonometry is the branch of mathematics that studies triangles and their parts — angles and sides. The fact that a simple three-sided figure became the object of a big field of study may surprise you at first, but your surprise disappears as you discover the practical value of trigonometry.

We did an informal survey to find the "pain and anguish" factors in math. You can be sure that trigonometry was high on the P&A list. It was also a big winner on the F&L (fear and loathing) list. It's one of the four great chillers in math (the others being algebra, geometry, and calculus — check out Chapters 12 and 14 for more on the first two), those classes that spark kids to stop paying attention.

Trigonometry (or *trig,* as you can freely call it) seems mysterious, but it's one of the most useful studies you get into. It's practical as well as conceptual, and it answers important questions in a fast and easy way. Plus, it's valuable in several professions, especially surveying, land engineering, architecture, and marine navigation.

In this chapter you see the basic ideas behind trigonometry. You also identify six functions and use them to do a little trig math.

Handling Triangles: More Angles than a Cornfield Maze

Trigonometry is the study of triangles, their angles, and their sides. It's especially about right triangles, but you extend it to apply to all triangles.

Look at what you probably already know about triangles (see Chapter 14 for more):

- A triangle has three sides.

- A triangle has three angles.

- The three angles in a triangle add up to 180 degrees. So if you know two angles, you get the third angle by subtracting those two figures from 180 degrees.

- With the Pythagorean theorem, you can find the length of one side of a right triangle when you know the length of the other two sides.

You're well on your way to handling triangles. But what about the rest? What happens when you know two angles and one side? Two sides and one angle? Can that get you anything? Yes it can, when you use trigonometry.

When you know three of a triangle's six parts (three sides and three angles), you can get all six parts, provided one of the given parts is the length of a side. That piece of magic is called *solving the triangle*, and it's brought to you by Trigonometrus the Magician, now appearing in this chapter.

And how does Trigonometrus pull a triangle out of his hat? Through an understanding of the relationships (or you can say ratios) between the sides. We discuss these relationships in the following section.

Here's 20 seconds of history. Before the Greeks, the Egyptians and Babylonians knew the ratios of the sides of a triangle, but they didn't know angles. (The Egyptians used a primitive slope measurement, the *seked,* to build the pyramids.) Indian trig concepts didn't get to medieval Islamic scholars until about the 10th century. Europe got on board in the 15th and 16th centuries. But they didn't "get it" until Nicolaus Copernicus (who proposed that the sun was the center of the solar system) wrote about it in Book II of his *"De revolutionibus orbium coelestium"* or *"On the Revolutions of the Heavenly Spheres."*

Did you study for the Triangle Measure test?

The word *trigonometry* comes from the Greek words *trigōnon,* meaning "triangle," and *metron,* meaning "measure." If you're speaking ancient Greek, this wording makes great sense. The problem for English speakers is that educators never translated the words into plain English. If they had, you'd now have a course called "Triangle Measure 101," which may cause less panic.

By Their Sines Shall Ye Know Them: Using Trigonometric Functions

Six trig functions describe the relationships between the sides of a triangle. Three are very important, and the other three are on the B team because they're *reciprocal functions*, defined as the reciprocals of the main functions.

The functions have long names and short abbreviations used in calculations. They are

Function	*Abbreviation*
Sine	sin
Cosine	cos
Tangent	tan

Reciprocal function	*Abbreviation*
Cosecant	csc
Secant	sec
Cotangent	cot

Now you know the names, but what do they mean? Figure 16-1 shows a triangle from which you can derive the functions. It's a right triangle, signified by the little square in angle C. It has angles A, B, and C. The sides are a, b, and c.

Figure 16-1:
Angles and sides of a triangle.

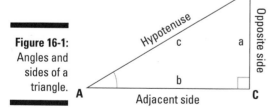

Angle A faces the side called the *opposite* side. It's next to the side called the *adjacent* side. The remaining side is the *hypotenuse*, which is always the longest side in a right triangle.

Examine the relationship between any two sides. The following sections discuss the three great relationships (ratios) and the three so-so relationships you get in any triangle.

Sine, cosine, and tangent: Three great relationships

The *sine* (abbreviated *sin*) is the ratio of the opposite side to the hypotenuse.

$$\sin A = \frac{\text{Opposite}}{\text{Hypotenuse}} = \frac{a}{c}$$

The *cosine* (abbreviated *cos*) is the ratio of the adjacent side to the hypotenuse.

$$\cos A = \frac{\text{Adjacent}}{\text{Hypotenuse}} = \frac{b}{c}$$

The *tangent* (abbreviated *tan*) is the ratio of the opposite side to the adjacent side.

$$\tan A = \frac{\text{Opposite}}{\text{Adjacent}} = \frac{a}{b}$$

At some point, as you see the value of knowing trig functions, you want an easy way to remember them. Try this mnemonic device:

1. **Start with what you know:**

   • **S**ine = **O**pposite ÷ **H**ypotenuse

   • **C**osine = **A**djacent ÷ **H**ypotenuse

   • **T**angent = **O**pposite ÷ **A**djacent

2. **Remember the first letter of each part of the formulas and combine them into a phrase.**

   We like "**S**ome **O**fficers **H**ave **C**urly **A**uburn **H**air **T**ill **O**ld **A**ge," but anything that works for you is fine. The letters SOH, CAH, and TOA each give you a key to the formula.

As a practical matter, you use other methods besides manual calculation to get an angle's trig values. You can also use trigonometry tables and calculators from the Internet, a scientific calculator, or the trig functions in a spreadsheet program.

Cosecant, secant, and cotangent: Three so-so relationships

The three less-common trig functions (cosecant, secant, and cotangent) are *reciprocal functions*. They're the three basic functions (see the preceding section) with the ratios inverted.

The *cosecant* (abbreviated *csc*) is a variation of the sine function. It's the ratio of the hypotenuse to the opposite side.

$$\csc A = \frac{\text{Hypotenuse}}{\text{Opposite}} = \frac{c}{a}$$

The *secant* (abbreviated *sec*) is a variation of the cosine function. It's the ratio of the hypotenuse to the adjacent side.

$$\sec A = \frac{\text{Hypotenuse}}{\text{Adjacent}} = \frac{c}{b}$$

The *cotangent* (abbreviated *cot*) is a variation of the tangent function, but with the terms inverted. It's the ratio of the adjacent side to the opposite side.

$$\cot A = \frac{\text{Adjacent}}{\text{Opposite}} = \frac{b}{a}$$

These function names are terrible, aren't they? Through the miracle of channeling, the ancient mathematicians want to apologize to you. Aryabhata discovered the sine and cosine. Muhammad ibn Mūsā al-Khwārizmī discovered the tangent. Abū al-Wafā' Būzjānī discovered the secant, cotangent, and cosecant. But the biggest apology comes from Albert Girard (1595–1632) because he was the first to use the abbreviations sin, cos, and tan in a paper. If he had called the functions "burger," "fries," and "shake," trig would be a lot more understandable today.

The law of sines

The *law of sines* shows Trigonometrus the Magician at his best. It states that

$$\frac{a}{\sin A} = \frac{b}{\sin B} = \frac{c}{\sin C}$$

This law applies to any triangle with sides a, b, and c and angles A, B, and C, where A is the angle opposite side a, B is the angle opposite side b, and C is the angle opposite side c.

The value of the law of sines is that when you know three measurements of a triangle (as long as they include at least one side), you solve for the fourth measurement with no problem. You must have two angles and a side or two sides and an angle.

Example: Surveying a River

You're at the Monongahela River in the Monongahela, Pennsylvania. How wide is the river? The only tools you have to for figuring the answer are your trusty surveying transit, a pad, and a calculator. Well, to make it a little easier, you also have a laptop with Wi-Fi and a spreadsheet program.

You identify trig problems by seeing whether you can construct a triangle. If you have enough information, you can solve for the answer. Your best trig friend is the law of sines (which we cover in the preceding section).

Use trigonometry. Establish a triangle, and then measure two angles and one distance (side). Figure 16-2 shows the setup. This triangle is a right triangle, as the little square in angle C shows. It has angles A, B, and C, and the sides are a, b, and c.

The mighty Monongahela

A
You are here
31°

c

b 500 feet

90°

Fort Defiance   Old Town

B   a   C

Figure 16-2:
Establish a
triangle to
determine
the river's
width.

1. **Starting at Old Town, sight Fort Defiance (which is also angle B) directly across the river from Old Town (which is also angle C) and then walk 500 feet north (at exactly 90 degrees) to A.**

2. Take a sighting of Fort Defiance from your new vantage point, A.

Angle A is 31 degrees.

3. Use the angles you know to calculate the third angle.

Because the three angles in any triangle add up to 180 degrees, and you know angles A (which you measured in Step 2) and C (which you established as a right angle), do the following subtraction to get angle B.

180 degrees – 90 degrees – 31 degrees = 59 degrees

Angle B is 59 degrees.

4. Use a calculator or a spreadsheet program to find the sines of angles A and B.

That works out to

sin A = 0.5150   sin B = 0.8572

5. Use the law of sines to find a, the distance across the river.

Use the terms that include a and b.

$$\frac{a}{\sin A} = \frac{b}{\sin B}$$

Substitute the values you know.

$$\frac{a}{0.5150} = \frac{500}{0.8572}$$

6. Use algebra to make the equation a little friendlier by multiplying both sides by 0.5150.

Your equation looks like the following:

$$a = \frac{500}{0.8572} \times 0.5150$$

The answer is 300.4317 feet.

Example: Locating a Wildfire

You're a U.S. Forest Service fire lookout on Bald Mountain, a couple of miles outside of Sterling City in Butte County, California. Like all Californians and forestry professionals, you take the threat of wildfire very seriously.

You see a plume of smoke and take a bearing, which shows the fire is 40 degrees east of north from your station. Bill Love, another lookout, is due east 20 miles away and sees the same plume; his bearing is 15 degrees east of north. You exchange information by phone. How far away is the fire from your tower?

As with the example in Figure 16-2 earlier in the chapter, establish a triangle and then measure two angles and one distance. Figure 16-3 shows the setup. This triangle is an obtuse triangle. It has angles A, B, and C. The sides are a, b, and c, but you're only interested in side b, the distance from your tower to the fire.

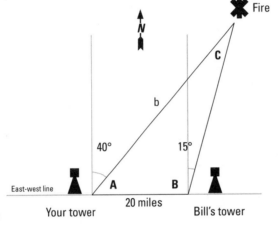

Figure 16-3:
Set up a
triangle to
figure the
distance to
the wildfire.

1. **Determine angle A by subtracting your 40-degree bearing from 90 degrees.**

   You use this method because your bearing was based on due north and the other tower is due east, creating a 90-degree angle. Your bearing and angle A total 90 degrees, so angle A is 50 degrees.

2. **Determine angle B by adding Bill's 15-degree bearing to 90 degrees.**

   You use this method because his bearing was based on due north and the east-west line between your two towers. That's a 90-degree angle. His bearing of 15 degrees plus the 90-degree angle total 105 degrees.

   Angle B is 105 degrees.

3. **Subtract to find angle C.**

   Because the three angles in any triangle add up to 180 degrees, and you know angles A and B, do the following subtraction to get angle C.

   180 degrees – 50 degrees – 105 degrees = 25 degrees

   Angle C is 25 degrees.

4. **Use a calculator or a spreadsheet program to find the sines of angles A and B.**

   The calculation works out to

   sin B = 0.26237 sin C = 0.13235

5. Use the law of sines to find b, the distance to the fire from your tower.

Use the terms that include b and c.

$$\frac{b}{\sin B} = \frac{c}{\sin C}$$

Substitute the values you know.

$$\frac{b}{0.26237} = \frac{20}{0.13235}$$

6. Use algebra to make the equation a little friendlier by multiplying both sides by 0.26237.

You end up with

$$b = \frac{20}{0.13235} \times 0.26237$$

The answer is 39.6479 miles.

This example is something of a classic and appears in various forms in various publications. The calculation is a fundamental one for the U.S. Forest Service and any agency where lookout towers are used.

Knowledge is power, except when it's not

If you're too clever (and maybe if you use too much trig), your creativity may get you in trouble on the job and in school. Jogger James Fixx (who helped start America's fitness revolution and wrote *Games for the Superintelligent* [Galahad Books]) gave this example in a Playboy Magazine interview:

Say someone gives you a barometer and asks you to use it to determine the height of a building. The person is probably expecting you to use an old formula about the difference in barometric pressure between ground level and the building's peak. Big yawn. Instead, you solve the problem in one of three different ways.

✔ You go to the top of the building and drop the barometer off. By timing its descent to the ground and knowing the formula for acceleration under gravity, you determine how high the building is. That's algebra.

✔ You stick the barometer in the ground at the point of the building's shadow, and by using the ratio of the barometer's shadow to the barometer's height and the building's shadow to the building's height, you calculate the height of the building. That's geometry.

✔ You go to the manager of the building and say, "Mr. Superintendent, I will give you this very fine barometer if you tell me how tall your building is." That's bargaining.

Such unorthodox thinking is bound to get you a head slap!

Part IV
Math for the Business of Your Work

The 5th Wave   By Rich Tennant

©RICHTENNANT

"Visionary architect or rotten mathematician, the jury's still out."

In this part . . .

People say that life is what happens to you while you're making other plans. That saying has a rough equivalent in work: "Your real job is never the same as the job described in the job description." New and unexpected experiences always cause change.

Part IV prepares you for three of the unscheduled, not-on-the-menu tasks of modern technical work. For example, you may be a great carpenter and a whiz at residential and commercial construction, but as you rise in the company, you may find yourself required to make presentations to clients. That's where the charts and graphs in Chapter 17 come in. Or perhaps you were hired to cut 2 x 4s, but now you're a supervisor who needs to check employee time cards and bill labor to clients. Never fear: Chapter 18's discussions of time math are here to help. And heaven help you if you have to buy the computers for the front office! Heaven may help eventually, but in the meantime, take a look at Chapter 19. It describes computer capacities and speeds.

Chapter 17

Graphs Are Novel and Charts Are Off the Chart

. . . .

In This Chapter

▶ Getting the most value out of graphs and charts

▶ Discovering the most important chart types

▶ Deciphering charts and graphs (and evaluating their quality)

▶ Creating great graphs

. . . .

*Y*ou've probably heard of graphic novels — stories told in pictures. Many people love them because pictures communicate so much (they're worth 1,000 words, after all).

Even if graphic novels aren't your thing, you've certainly heard of and used charts — if you've ever used a road map (a visual representation of geography and distance), you've used a kind of chart.

Graphs and charts are the mathematical equivalents of these everyday items. They may not be as exciting as manga (the picture stories so popular in Japan), but mathematical charts and graphs easily explain information and concepts and help you effectively communicate ideas to other people.

In this chapter, you discover the types of graphs and their parts and become an expert in reading and making them.

Defining Charts and Graphs and Their Advantages

In practical math, a *graph* and a *chart* are the same thing. They both refer to a graphic that shows tabular data in a visual form, otherwise known as an *information graphic*. You can pretty much use the words *chart* and *graph* interchangeably.

Figure 17-1 shows some typical types of charts and graphs. Don't worry yet about what the types are.

Figure 17-1:
Types of
charts and
graphs.

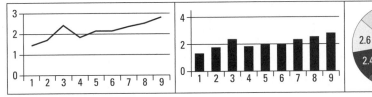

If you want to get formal about definitions, one source says that "In mathematics, a graph is an abstract representation of a set of objects where some pairs of the objects are connected by links."

Here are some advantages of using and making charts and graphs:

- ✔ You can understand the numbers in charts more quickly and easily because a chart is a visual (not textual) presentation. Trying to interpret a column or table of numbers may make your eyes roll up into your head, but a good chart or graph does some of the interpreting for you, making the information easier to get. Similarly, when you make charts and graphs, you communicate information better.

- ✔ You can easily understand large quantities of data and see the relationships among different items.

The advantages apply to most trades, especially the business side of any trade. You can use charts to compare the properties of materials, the cost of materials, or the rate you're using materials up.

One chart disadvantage: Charts can be less precise than tables or listed information. Charts and graphs provide easy-to-read visuals, but they can't always concretely portray the more minute specifics of data.

Paying Tables Their Proper Respect

Tabular data is information that appears in a table of rows and columns. Although we champion charts and graphs over tables in the preceding section, we can't dis tabular information entirely — it's the basis for graphs. (If you build charts in Microsoft Excel, you start with rows and columns: tabular data.) Tables may be a little hard to figure out, but they're the form that specifications come in, including the materials you work with every day, so you have to know how to deal with them. And no matter what, a table on its worst day is better than the same text *inline* (written out) on its best day.

For those working in cooking or pastry, a recipe is a table of data. Welders compare welding rods by using tables, and machinists use tables when they look up feed rates. And lots of folks buy computers for business or home by reading endless tables of specifications. Some online vendors build the tables for you with a "compare side by side" feature.

For example, compare the following text describing quantities of laboratory reagents with the same information shown in table form in Table 17-1:

> "We have on hand 2.5 kilograms of sodium hydroxide, 2.5 liters of sulfuric acid, 500 milliliters of nitric acid, and 500 milliliters of glacial acetic acid."

Now look at Table 17-1.

Table 17-1	Inventory of Reagents in Storeroom
Reagent	*Quantity*
Sodium Hydroxide	2.5 kg
Sulfuric Acid	2.5 l
Nitric Acid	500 ml
Acetic Acid, Glacial	500 ml

The table is orderly and somewhat visual, which makes finding information in the table easier than in the inline text.

Introducing the Three Most Important Types of Charts

A chart can take many forms because certain types of charts show information better than other types. The three most important types of charts (the ones you encounter most often in your work) are line graphs, bar graphs, and pie charts. These are the charts you're most likely to construct when you want to communicate with other people, so they're the ones we discuss in the following sections.

In precise mathematical terms, there are a bazillion types of charts. Some are variants of the three types in this chapter. A few examples include area charts, scatter charts, open-high-low-close charts, cartograms, timeline charts, pedigree charts, bubble charts, Gantt charts, PERT charts, Pareto charts, organization charts, and flowcharts.

Walking the line graph

A *line graph* contains lines made up of connected *data points*. The graph often shows changes in a quantity or the value of something over time. Figure 17-2 is a typical line graph.

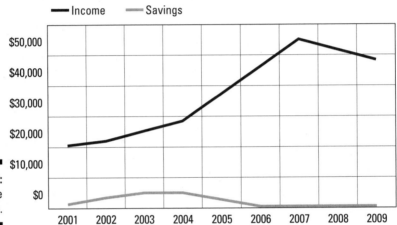

Figure 17-2:
Typical line
graph.

The graph shows a trend in income and savings over a period of time (2001 to 2009). From this graph, you quickly see trends. For example, income rises over the first seven years and then starts to decline in the last two (as the hard times hit). Savings start to grow but peak in about 2004 — and have declined ever since.

Sidling up to the bar graph

A bar graph isn't a map to the local bars. A *bar graph* or *bar chart* uses bars (either horizontal or vertical) of various lengths to show values. Most bar graphs show information for only one item per bar.

Figure 17-3 shows a vertical bar graph.

The numbers in Figure 17-3 are real percentages based on a *U.S. News & World Report* article written by Deborah Kotz for the March 28, 2008, issue.

Share of U.S. Babies Delivered by C-Section

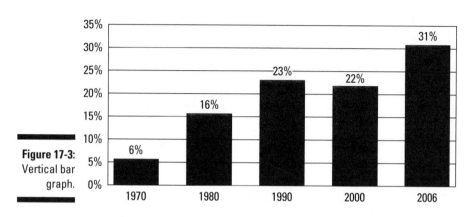

Figure 17-3:
Vertical bar
graph.

Figure 17-4 shows a horizontal bar graph based on figures from the 2006 CIA World Factbook.

Inflation Rates in Various Countries (%)

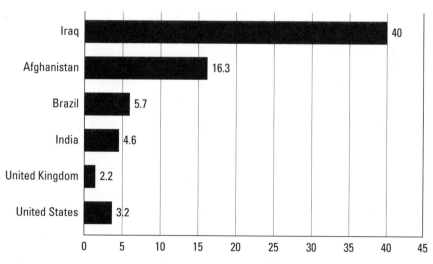

Figure 17-4:
Horizontal
bar graph.

A *histogram* is a special kind of bar chart. It uses bars to show groups or bands of data, such as the average height of students in various age groups or the percentage of various age groups in a city's population, as bars.

The technique for calculating the values is called *binning* because the data are put into groups or bins. Figure 17-5 is a histogram.

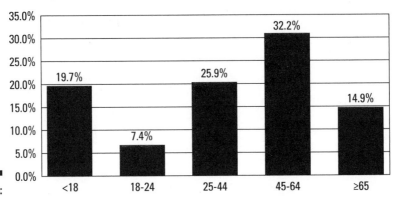

Population in Nevada City, California

2000 U.S. Census–Nevada City Population: 3201

Figure 17-5: A histogram.

Histograms are very useful, but read (and create) them carefully. They're prone to distortion if the creator calculates the average values poorly or if the value ranges of the bins aren't equal. In Figure 18-5, for example, only two of the middle bins have equal ranges. Similarly, the bins for the Nielsen TV ratings age demographics are all over the place because advertisers want to sell to some very specific (and not equal) age bands.

Getting a piece of the pie chart

As you can see in the preceding sections, a line graph is a fine graph, and a bar chart is a good start. However, you may find that you want to make the pie chart "*my* chart." Seriously, the pie chart is probably the most widely-used chart.

The *pie chart* or *pie graph* represents information as a circle divided into sectors or slices. The different-sized sectors show the relationship among the items of information. The pie chart is a super type of graph when you want to show how parts of something add up to 100 percent of a whole thing.

The earliest known pie chart is credited to William Playfair's *Statistical Breviary* of 1801. And who, pray tell, is William Playfair? For one thing, he's a guy with such a great name that you shouldn't finish this book without knowing it. He was also a Scottish engineer and economist credited with pioneering statistic graphics and inventing the line graph and bar chart of economic data, as well as the pie chart and circle graph.

Figure 17-6 illustrates two pie charts. The first one is a regular chart, and the second one is *exploded* to emphasize a particular wedge.

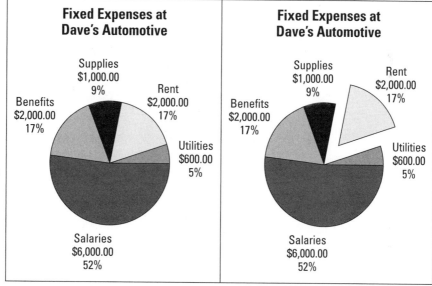

Figure 17-6:
A regular pie chart and an exploded pie chart.

Reading Charts and Graphs (And Recognizing a Bad One)

You do two things to get the most out of a chart. First, you recognize its parts, and second, you draw meaningful conclusions from the information presented. A good chart pays you back by communicating a lot of good information. Unfortunately, not all charts are good charts; some charts may be incomplete or distorted. The following sections show you how to identify good charts as well as the parts that make them up.

For a start, the parts of a chart

The first step in reading a chart is to figure out which standard chart parts it does and doesn't have. Here's a short list of the parts you can expect on most charts. Not all charts have all parts, but all charts have at least some of them.

- ✔ **Title:** A chart's *title* is usually at the top or bottom of the chart. It tells you what the chart is about.

- ✔ **Axis:** A line graph or bar graph has two *axes* — a *horizontal axis* and a *vertical axis* — that cross at a point called the *origin*. Each point in the graph is some distance to the left or right of the vertical axis and above or below the horizontal axis. ***Note:*** In many graphs, the plot area (see the later bullet) shows only one region of the axes. For example, the charts in this chapter show only the area to the right of the vertical axis and above the horizontal axis.

- ✔ **Scales:** The chart's *scale* typically shows the units the information is presented in. The units may be time, numbers, percentages, and so on. The horizontal and vertical axes each have a scale, and each should have some sort of periodic graduations.

Expect each scale to have a *label*. A scale with units (for example, 50, 150, 200, 250, and 300) isn't very useful if it doesn't tell you what they represent. With these sample numbers, a label such as "Miles Driven Each Week" is a necessary description.

Even in a pie chart, expect to find units clearly described.

- ✔ **Plot area or data area:** This area is where the information appears. Basically, it's the area that isn't the title, axis, scales, or labels.

- ✔ **Grid:** Some charts have a *grid,* a series of vertical or horizontal lines that help you easily see and compare the values. Expect to see grid lines on a line graph or a bar graph but not on a pie chart.

- ✔ **Data:** The information in a graph can be dots, lines, bars, pie wedges, and symbols. These are *data points*. Truly, you can plot the data any number of ways as long as the method communicates the information well.

Sometimes you see more than one variable on a line graph or a bar graph. This setup allows you to compare multiple quantities at the same time.

- ✔ **Symbols:** A *symbol* is a picture used to show the amount of an item. It makes a more interesting chart, but be sure to label what the symbol represents and the value of each symbol. Figure 17-7 is an example of a chart that uses symbols.

✔ **Legend:** The chart may include a *legend,* a list of the variables in the chart that describes what each color or pattern of the lines or bars in the graph represents. For example, weight and height may appear on a line graph showing a child's growth over time. Each line may be a separate color or be solid or dashed so you can tell them apart. In a pie graph, the legend describes what the color or pattern of each slice represents.

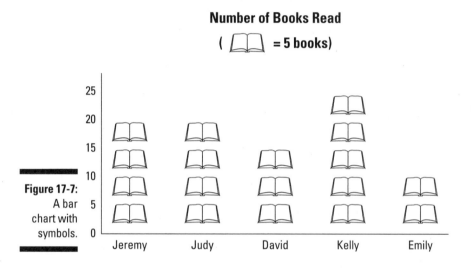

Figure 17-7:
A bar chart with symbols.

The good, the bad, the ugly, and the inaccurate

Edward Rolf Tufte (pronounced *tuf*-tee) is famous in the world of visual design as the supreme authority on visual communication of information. If you see a good, professionally constructed graph (for example, online in a major newspaper), the designer was probably influenced by Tufte. If you see a bad graph (and there are quite a few), there's no telling what the designer was under the influence of.

Tufte invented the term *chartjunk* to describe the stuff on graphs that's useless or obscures the meaning. Based on his books and common sense, we present four categories of charts: good, bad, ugly, and inaccurate.

✔ **Good graphs:** A good graph presents good data in the best format. It has every needed element, but only those elements it truly needs, making it complete and accurate. A good graph isn't confusing in any respect.

✔ **Bad graphs:** A bad graph is pretty much the opposite of a good one. A bad graph may present the information in the wrong format (for example, pie chart instead line graph) or have missing elements (such as units) or poorly chosen units. Omitted or incomplete information makes a bad graph even worse.

✔ **Ugly graphs:** An ugly graph has too much of everything. It may be accurate, but nobody cares because nobody can read it. Ugly graphs are the ones with too much information, too many labels, too many variables, or too many colors. In addition, the type size may be too small, and all the data may be too crowded.

✔ **Inaccurate graphs:** The worst graphs are those whose representations (especially bars and symbols) make the variables look bigger or smaller than they are. Another contributor is out-and-out wrong data. Such graphs are inherently inaccurate or present information in a distorted, inaccurate way. Omitting data can also make a chart inaccurate.

The worst inaccuracies kill people. When NASA was debating whether to launch the space shuttle *Challenger*, the managers were presented with badly prepared charts from the manufacturer of the shuttle's solid rocket-fuel boosters. The charts left out information, presented information badly, included unimportant information, and failed to put related pieces of information on the same chart. NASA decided to launch, the O-rings failed, and the crew perished.

Making Charts and Graphs

Graphs are great communication tools if you know how to make them properly. When you make your own charts and graphs, you become a graphic designer as well as a practical mathematician. The basic "good graph" rules from the preceding section apply to graphs you construct yourself:

✔ **Use the proper type of chart for the information you're presenting.** The wrong kind of chart can't communicate your information as clearly the right chart can.

✔ **Use only essential data.** Avoid putting useless information in your work. Excess information confuses the reader, makes reading harder, and shortens the reader's attention span.

✔ **Use accurate data.**

✔ **Use good, clear, and meaningful titles, legends, and labels.** That includes good fonts in sizes that aren't too big or too small. Make sure your title clearly and accurately represents the chart's content.

✔ **Use appropriate units.** Measuring in inches when feet are more intuitive makes your chart harder to understand. Also, avoid mixing U.S. American units with metric units.

✔ **Use a plain white background that allows the important information to stand out.** Decorative or fancy backgrounds may make your graph harder to read.

If you use Microsoft Excel, try the chart wizards. The software is far from perfect, but it can handle of lot of the tedious tasks in chart-making. Just be cautious (what you get out is only as good as what you put in), and don't get fancy; you may end up with a chart that's too cutesy.

The following sections give you instructions for creating solid, useful line, bar, and pie graphs.

Creating line graphs

The following is a simple procedure for making a good line graph:

1. **Gather all your information first.**

   Keep everything neat and orderly.

2. **If you're making a line graph by hand, use *graph paper* (the paper with vertical and horizontal lines).**

   If graph paper isn't available, a piece of plain paper works fine.

3. **Make your graph large.**

   Use the paper to its fullest. It helps prevent cramming information into tight spots.

4. **Draw scales for both the vertical and horizontal axes and add in the marks for units.**

5. **Label both axes.**

   Always name the units and what information they represent. For example, you may write "Fuel Consumption (gallons)."

6. **Plot your points and connect them with a line.**

   Use dots, small circles, small diamonds, or small triangles. You can use other symbols, but don't get too cute.

7. **Add a clear title.**

8. **If you have multiple variables, add a legend.**

Building bar graphs

Like line graphs (see the preceding section), you use a simple procedure for making a good bar graph.

1. **Gather your information.**

2. **Use graph paper.**

3. **Make your graph large.**

4. **Draw a scale for both the vertical and horizontal axes and add in the marks for units.**

5. **Label both axes.**

   Remember to include both the units and what they represent.

6. **Draw the bars.**

7. **Add a title.**

8. **If you have multiple variables, add a legend.**

Putting together pie charts

In many ways, making a pie chart is very similar to making line and bar graphs (see the preceding sections). One difference: You don't need to worry about graph paper — plain paper is fine. Follow this procedure:

1. **Gather your information.**

2. **Use a compass to draw the circle and a protractor to draw slices that are about the right size.**

   Remember to make your graph large. If you have a slice that represents 15 percent, you don't necessarily have to measure out precisely 15 percent of the circle; just be sure it doesn't obviously look like 25 percent.

   When you have too many categories to make a readable pie chart, consider combining the items. For example, in a housing-costs scenario, you can combine plumbing fixtures with plumbing connection work. You lose precision, but remember, a graph isn't a precision instrument. It's supposed to provide a picture.

3. **Use color to clearly differentiate your slices.**

   Pie charts are one instance where color is a really constructive element in a chart. Using colors in a pie chart makes separating each category much easier. In a pie chart, you're always trying to separate the data as much as compare it.

   4. Add a title.

If you're using two pie charts side by side with the same categories of data, be consistent. For example, if you have pie charts to show budgets for two business operations, use the same size pies, the same data elements, the same labels, and the same colors for the corresponding data elements. Yes, the sizes of the slices are different in each graph, but the colors help the reader's eyes distinguish and compare them.

Some people will tell you that pie charts aren't good for comparing two sets of data, and they're frequently right. If you find yourself using two pie charts side by side, be careful. It doesn't automatically mean you've made bad choice, but you've got to be certain the presentation best conveys what you want to say.

Additionally, pie charts are susceptible to a couple of unique traps. Pie chart errors include

- ✔ **Having too many slices:** Excessive slices are a prime cause of confusing pie graphs. They indicate that you probably picked the wrong type of graph for the information presented. Consider using a bar graph instead.

- ✔ **Failure to label the slices correctly:** If you're trying to tell your reader that the food bill for your kitchen operation is 25 percent of your total budget and your label points to the 10 percent allocated for electricity, you've got an inaccurate chart.

Try the Microsoft Excel chart wizards. Be careful if you choose a 3-D pie chart — sometimes this type of presentation distorts the data by making some pieces look a little larger than they really are.

Example: Tracking Weight and Height in a Pediatric Practice

You work in a pediatric practice and regularly measure the height and weight of your young patients. When they were babies, you measured their length. (My, they grow up so fast!)

Young John Doe is now 15. The doctor tells you his height is about right for a boy his age, but he's looking a little porky. She asks you to prepare a graph showing John's height and weight over his childhood years. She gives you the information in the form of Table 17-2.

Table 17-2	John's Height and Weight from Ages 8 to 15	
Age	*Height*	*Weight*
8 years old	45 in	57 lbs
9 years old	49 in	61 lbs
10 years old	51 in	70 lbs
11 years old	52 in	77 lbs
12 years old	58 in	85 lbs
13 years old	60 in	100 lbs
14 years old	63 in	115 lbs
15 years old	66 in	125 lbs

Create the chart:

1. **Select the best chart type.**

   A line chart is excellent for showing change over time. You use one line for weight and another for height

2. **Gather the information.**

   This one's easy: The doctor gave you the info in the table.

3. **Draw a scale for the horizontal axis, which shows John's ages.**

4. **Draw a scale for the vertical axis.**

   The vertical axis can show both height in inches and weight in pounds. Your legend (Step 6) explains what the data points mean.

5. **Plot your points (use dots, small circles, small diamonds, or small triangles) and connect them with a line.**

6. **Because you have multiple variables, add a legend.**

   Use "Height (inches)" and "Weight (pounds)," which clearly describe your info.

7. **Add a clear title.**

   "John Doe Height and Weight" clearly indicates what your chart presents.

Your line graph should look something like Figure 17-8.

Comparison with published averages shows that John's increases in height and weight are entirely average. If John is looking like a little doughboy, he's probably spending too much time in front of the television. Prescription: exercise.

John Doe Height and Weight

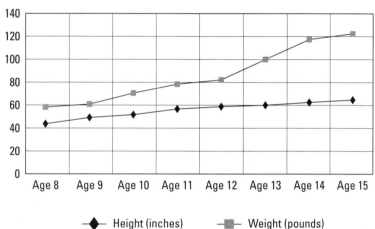

Figure 17-8:
A line graph showing John Doe's height and weight over the years.

Example: Cost of Materials in Residential Construction

You're rising fast in a residential real estate development company. For better or worse, your job now requires a lot more than swinging a hammer. Your boss has asked you to create a pie chart that provides a good visual that shows what goes into building a residence.

The good news is that your company's typical house is 1,600 square feet with an attached garage. It's a six-cornered (L-shaped) design. Further, it has no fireplace, woodstove, or attic, which keeps costs down. The bad news is that your company builds in the Sacramento suburbs in Northern California. Even a modest house costs a lot to build there.

You use an Internet calculator to get the costs for 28 separate cost categories, with a total cost of $220,769. (People in other parts of the country may now recover from fainting and continue reading.)

As a simple example, use the six *costliest* categories for your pie chart (after clearing it with your boss, of course). These categories reflect about half the cost of building a house and are shown in Table 17-3.

Table 17-3	Costliest Categories in Building a House			
Item Name	**Material**	**Labor**	**Equipment**	**Total**
Rough Carpentry	19,825	27,177	--	47,002
Exterior Finish	10,555	5,853	785	17,193
Foundation, Piers, Flatwork	6,131	8,955	1,519	16,605
Roofing, Flashing, Fascia	8,289	6,456	--	14,745
Interior Wall Finish	5,281	7,573	--	12,854
Plumbing Rough-in/ Connection	3,371	7,603	492	11,466

Create your chart:

1. **Select the chart type.**

   In this case, it has to be a pie chart, because that's what your boss asked you to prepare. But it's also the best way to show the parts of a whole.

2. **Gather the information.**

   This information's in the table. However, you judge that these tasks are better expressed as percentages, so convert the total cost for a task into a percentage (total cost for a task, divided by total cost for the house, multiplied by 100).

3. **Use a compass to create the circle.**

4. **Draw the wedges with a protractor.**

5. **Add a legend or label individual slices of the pie.**

6. **Add a clear title, such as "Cost of Building a House."**

Keep in mind that this example gives the details on only 6 of the 28 categories. To keep the chart simple, you combine the remaining 22 categories into an "All Other Costs" category.

Your pie chart should look a lot like Figure 17-9.

Cost of Building a House
(Shown as Percents)

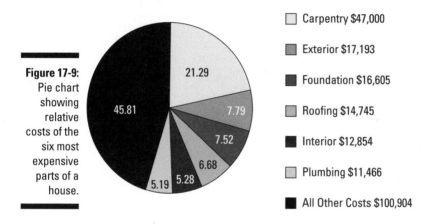

Figure 17-9:
Pie chart
showing
relative
costs of the
six most
expensive
parts of a
house.

☐ Carpentry $47,000

▨ Exterior $17,193

■ Foundation $16,605

▨ Roofing $14,745

■ Interior $12,854

☐ Plumbing $11,466

■ All Other Costs $100,904

Chapter 18

Hold on a Second: Time Math

Time is an essential and undeniable part of the human experience. That's a lofty statement in a book about technical mathematics. But it's true — reckoning time is a vital component of almost every civilization.

Time is easy to read, report, measure, and calculate. In this chapter, you see how to denote time in both common and uncommon ways. You convert time into traditional, fractional, and decimal forms and perform some time math.

Many societies are run by time: They're deadline-driven and have a very low tolerance for tardiness. But at least one place *has no concept of time*. The Pirahã people on the Mai ci river in the Amazon rainforest speak by humming and whistling; they have no more than eight consonants and three vowels and have no numbers and no time.

Dividing Time into Hours, Minutes, and Seconds

Time is the most precisely measured quantity on earth. Atomic clocks are accurate to 10^{-9} seconds (one *billionth* of a second) per day, and accurate time is essential for maintaining the frequency of television broadcasts and operating the GPS system.

Time is important in both business and personal activity in the following ways:

- ✔ **Time is a measure of progress or capacity.** For example, you say "How much did you get done on the project today?" or "How many blood samples can you analyze in an hour?"

- ✔ **Time marks the point at which something will happen.** For example, you say "When does the store open?" or "When does your flight leave?" (You may not believe it, but commercial airplanes are supposed to take off and land "on time.")

- ✔ **Time is the basis for good manners and cooperation.** You're expected to arrive at work on time. Also, being "fashionably late" to meet a friend is a sign of discourtesy.

If your career requires delivering anything, cooking food, curing concrete, centrifuging in the lab, meeting people or banking, time is a factor. If you bill clients for time expended, or if you're paid by the hour, time is the key to business and personal income. Many activities in business and in personal life run "on the clock."

To make time easier to quantify, it's divided into units: hours, minutes, and seconds.

A day is made up of 24 hours. An *hour* contains 60 minutes, and a *minute* contains 60 *seconds*. Another way to think of it is to say a second is 1/60 of a minute, a minute is 1/60 of an hour, and an hour is 1/24 of a day.

Timekeeping is the act of determining the time. You call it telling time. *Tell* means count, which is why the person at your bank is called a teller. *Duration* or *elapsed time* is the amount of time that passes between two events. For example, a race has a start time and an end time. The amount of time it takes the winner to cross the finish line is the duration.

Ancient civilizations, including Egypt, Babylon, Sumer, the Indus Valley civilization, and China, divided the daylight time into 12 parts. One source credits the Egyptians for dividing the night into 12 parts.

Here's a reminder about some things you already know (and if you didn't know these things, keep quiet):

- ✔ Larger units of time are the day, week, month, year, decade (10 years) century (100 years), and millennium (1,000 years).

- ✔ Smaller units of time are the tenth of a second (0.1 seconds), the hundredth of a second (0.01 seconds), the millisecond (0.001 seconds), the microsecond (a millionth of a second or 0.000001 seconds) and the nanosecond (a billionth of a second, or 0.000000001 seconds).

What is time, exactly?

So what is time? Everyone knows that it exists, but it seems that no one can define it precisely.

✔ For science and everyday life, time is a system for measuring the occurrence of events and the duration between events. The definition is circular: Time is a system for measuring time.

✔ Religion, philosophy, and science all study time but have disagreements and controversy about it among them.

✔ Time is objective, except that it's not. The general theory of relativity says that the clock in a fast-moving spaceship ticks more slowly the one on earth. That's called *time dilation*.

✔ Time is real, except when it's not. *Imaginary time* is a concept in quantum mechanics. Stephen Hawking discusses it in *A Brief History of Time* (Bantam).

✔ Time is infinite and eternal, except when it's not. Love songs tell you time is eternal (sometimes singing of "forever" and sometimes "till the end of time"), but speculative physics suggests that time started with the Big Bang, and the Abrahamic faiths believe that time started with the Creation.

Is your head still in one piece? Writing about time can make a very long book, but we don't have the time.

Not everyone uses the same conventions, styles, and standards for writing and speaking the time and the date. For example:

✔ In most countries, you write the last day of 2014 as 2014-12-31. The last second of the year is 23:59:59.

✔ In the United States, you use America's traditional notation — you write that date as 12/31/2014 and that time as 12:59:59 p.m.

There's a Time for Us, Somewhere a Time for Us: Time Notation Systems

Methods of showing and saying the time have varied over thousands of years. Today, two methods are most prominent; we cover these two methods, and a few more that are commonly used in some circumstances, in the following sections.

12-hour notation

If you live in the United States, you're familiar with the 12-hour clock. In both analog and digital form, it's the clock you typically use to tell time. Figure 18-1 shows both analog and digital 12-hour clocks.

Figure 18-1:
Two 12-hour clocks, analog and digital.

2:23 PM

The 12-hour analog clock has numbers from 1 to 12 on the face. It's understood that the 12 hours represent either "a.m." or "p.m." It also has two hands, an *hour hand* (the short hand) and a *minute hand* (the long hand). On many clocks, the minute hand lines up with one of many small marks between the big numbers. Sometimes there's a third, sweeping hand called a *second hand*.

After 12:59, you start counting minutes from 1:00, so the minute after 12:59 p.m. is 1:00 p.m. To add to the confusion, the last minute of morning is 11:59 a.m. The first minute of afternoon isn't 1:00 p.m.; it's 12:00 p.m. (which is noon, not midnight, even though the *p.m.* may suggest nighttime).

a.m. comes from the Latin phrase *ante meridiem* and means "before midday." *p.m.* is from the Latin word phrase *post meridiem* and means "after midday."

It's traditional. It's American. It also makes telling time and time calculations harder for two reasons:

✔ You have to figure out whether it's "a.m." or "p.m."

✔ You have to borrow or carry 12 a lot in time calculations.

24-hour notation

If you live outside the United States, work in aviation, or serve in the military, you more likely you use a 24-hour clock rather than a 12-hour clock (see the preceding section) to tell time. Figure 18-2 shows a 24-hour clock in both analog and digital form.

Figure 18-2:
Two 24-hour clocks, analog and digital.

The 24-hour analog clock has numbers from 1 to 24 on its face. Each number is a distinct hour in the 24-hour day. Like the 12-hour analog clock, the 24-hour analog clock has an hour hand and a minute hand. The minute hand usually lines up with one of many small marks between the big numbers. Like the 12-hour analog clock, the 24-hour analog clock frequently has a second hand. Also, there's no need to designate "a.m." or "p.m." on a 24-hour digital clock.

After midnight (00:00), you start counting time from 0. The minute after midnight is 00:01 and the minute after 00:59 is 1:00. After the noon hour (12:00), the counting continues in the same way, so the minute after 12:59 is 13:00. The last minute of the day is 23:59.

Military time is the way you express time in the U.S. military and is an example of 24-hour notation in action. Here are some examples of how to say military time:

- ✔ If your work starts at 8:00 a.m., you start at "Oh [as in zero] eight hundred hours."

- ✔ An 11:00 a.m. meeting starts at "Eleven hundred hours."

- ✔ If your work ends at 5:30 p.m. on the 12-hour clock, it's 17:30 on the 24-hour clock. You say "Seventeen thirty hours."

- ✔ If an activity begins at midnight, you say "24 hundred hours."

Greenwich mean time (GMT)

Greenwich mean time (or GMT) is a world standard for time. GMT divides the world into 24 time zones. Each time zone is relative to the Royal Observatory at Greenwich (*gren*-ich), England, and covers about 15 longitudinal degrees of the earth's circumference. Each time zone uses three letters to designate it. In aviation, GMT is practically identical to Zulu time (but we discuss that further in the following section).

In the continental United States, the times zones are eastern time (EST or EDT, depending on whether daylight saving time is in effect), central time (CST or CDT), mountain time (MST or MDT), and Pacific time (PST or PDT).

Greenwich mean time (or GMT) originally represented *mean solar time*, which was an approximation of true solar time. Neither mean solar time nor true solar time are accurate, because they don't account for seasons of the year, the angles at which sunlight hits the earth, the gradual speeding up of the moon and the gradual slowing down of the earth. GMT has been replaced in all but name by Universal Time Coordinated (UTC — see the following section).

If you have a PC running Microsoft Windows, you can see all the time zones by clicking Start→Control Panel→Date and Time→Time Zone. All the time zones are there.

You express all the time zones as positive or negative "offsets" from GMT. For example, you express the time zone for Cairo or Jerusalem as GMT+2, UTC+2 or GMT/UTC+2.

You can see your offset from GMT in your e-mails. In Mozilla Thunderbird, select a message and click View→Message Source. In Microsoft Outlook, select a message, right click, and then click Message Options. The box that pops up shows you the message header, including the offset. Here is a message header showing an offset of GMT–8:

Mon, 18 Jan 2010 10:31:53 –0800

Because Greenwich is the center of things, the home of the *prime meridian,* its longitude is 0° 0' 0". Because each time zone accounts for 15 degrees of the earth's circumference, a person who's six time zones away from Greenwich (in, say, Chicago) is about 90 degrees or one quarter around the world from England.

UTC and Zulu time

Universal Time Coordinated is also known as *UTC, Universal Time,* and *atomic time.* By international agreement, UTC replaced GMT on January 1, 1972. Most people still use the term GMT.

Zulu time is another term for GMT/UTC. It's used universally in aviation because it helps minimize confusion when flying between time zones. For example, the phrase, "Land the plane at 3:00 pm," is meaningless in international aviation, as it could refer to the departure airport's time zone or the arrival airport's time zone. Either of these might be following standard time or daylight saving time. Instead, all pilots use the same 24 hour clock.

Civilian time zone designations are three letters in length (for example, EST, CST, MST, and PST), while Zulu time zones have one-letter alphabetical designations. The follow example shows several offsets and a city in each time zone.

Offset	Letter	Name	City
GMT+0	Z	Zulu	London
GMT+1	A	Alpha	Berlin
GMT+2	B	Bravo	Helsinki
GMT+3	C	Charlie	Baghdad

The time zone designations are taken from the NATO phonetic codes. The system doesn't use the letter *J,* and the letters *M* and *Y* are special because they designate the area around the International Date Line: international time zone east (UTC+12) and international time zone west (UTC–12).

Zulu time isn't named after the Zulu, the largest ethnic group in South Africa. It's named after the time zone where the Greenwich Observatory is located.

Swahili time

Swahili time runs from dawn to dusk. If the day starts at what you'd call 6:00 a.m., 7:00 a.m. is 1:00 in Swahili time. Figure 18-3 shows a 12-hour Swahili clock.

Figure 18-3:
A clock showing Swahili time.

saa nne na nusu asubuhi
"hour four and a half morning"
(10:30 a.m.)

Here's a quick lesson in Swahili timekeeping:

Swahili	*English*	*Time*
saa moja asubuhi	hour one morning	7:00 a.m.
saa tisa usiku	hour nine night	3:00 a.m.
saa mbili usiku	hour two night	8:00 p.m.

Swahili is a Bantu language. It's the *lingua franca* (unofficial shared language) of much of east Africa and the national language of four countries. It's also the main language of Tanzania (where there are 126 languages), even though English is the official language for school and the high courts.

Bible time

The Christian Bible treats the time of day in much the same way as Swahili time. The day starts at what you'd call 6:00 a.m. There are no clocks in the Bible, but modern advocates of a Bible clock place a 1 for the first hour at the start of the day.

Consider the parable of the workers in the vineyard (also called the parable of the laborers in the vineyard or the parable of the generous employer) from Matthew 20:1–15.

> 1 For the kingdom of heaven is like a householder who went out early in the morning to hire laborers for his vineyard.
>
> 2 After agreeing with the laborers for a denarius a day, he sent them into his vineyard.
>
> 3 And going out about the third hour he saw others standing idle in the market place;
>
> 4 and to them he said, "You go into the vineyard too, and whatever is right I will give you." So they went.
>
> 5 Going out again about the sixth hour and the ninth hour, he did the same.

Figure 18-4 shows the time of day the employer sent laborers into the vineyard, using a Bible clock.

Stardates

Sorry to say it, but stardates are completely fictional. Executive Producer Gene Roddenberry invented them for the original 1966 *Star Trek* television series. Officially, a *stardate* is a decimal number rounded to a single decimal place and used as a means of specifying absolute dates in the *Star Trek* universe. Unofficially, a stardate is a literary device designed to keep the screenwriter out of trouble. The idea behind stardates was to avoid mentioning the century *Star Trek* was set in (about 200 years in the future) and to avoid arguments about the plausibility of some of the technology.

Originally, to make a stardate, you picked any combination of four numbers and added one digit after the decimal place. The day is divided into ten units with noon at the middle of the stardate, so, for example, 1421.5 is noon on a particular day, and 1422.5 is noon the next day. Stardates only have to be consistent within an episode.

You can find several stardate calculators online. Why more than one? Because the original series, its offshoots *Star Trek: The Next Generation* and *Star Trek: Deep Space 9,* and the *Star Trek* movies aren't in exact agreement.

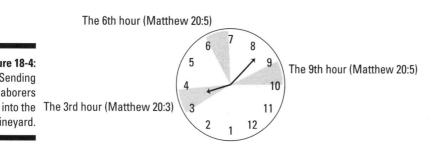

The 6th hour (Matthew 20:5)

The 9th hour (Matthew 20:5)

The 3rd hour (Matthew 20:3)

Figure 18-4: Sending laborers into the vineyard.

Converting Time

Time conversions aren't difficult. You just need multiplication and division (both of which we cover in Chapter 5). You use the following conversion factors:

- ✔ **Minutes to seconds:** 1 minute = 60 seconds
- ✔ **Seconds to minutes:** 60 seconds = 1 minute
- ✔ **Hours to minutes:** 1 hour = 60 minutes
- ✔ **Minutes to hours:** 60 minutes = 1 hour

You write hours and minutes as hh:mm, where *hh* represents the hours and *mm* the minutes. They're separated by a colon (:). For example, 1:15 is 1 hour and 15 minutes. You write minutes and seconds as mm:ss; *mm* still stands for "minutes" and *ss* represents the seconds. For example, 12:13 is 12 minutes and 13 seconds. This notation can be easily confused with the hour-minute notation, but your problem will likely identify (or give you the info to identify) which units you're dealing with. When you have hours, minutes, and seconds, write the expression as hh:mm:ss. For example, 1:17:45 is 1 hour, 17 minutes, and 45 seconds.

Going from minutes to seconds and back again

When you have a time with only minutes, you can easily convert it to seconds by multiplying by 60 (because a minute contains 60 seconds). For example, to convert 3 minutes, the equation is

$$3 \text{ minutes} \times \frac{60 \text{ seconds}}{1 \text{ minute}} = 180 \text{ seconds}$$

When you have minutes *and* seconds to convert, multiply the minutes by 60 and add the remaining seconds. A sample equation (using 3:15) is

$$3:15 = 3 \text{ minutes and 15 seconds}$$

$$3 \text{ minutes} \times \frac{60 \text{ seconds}}{1 \text{ minute}} = 180 \text{ seconds with 15 leftover seconds}$$

$$180 \text{ seconds} + 15 \text{ seconds} = 195 \text{ seconds}$$

To go the other way and convert seconds to minutes, simply divide your number of seconds by 60 to get minutes. If you have seconds that are evenly divisible by 60 (using an example of 180 seconds), this division is really easy:

$$180 \text{ seconds} \div 60 \text{ seconds/minute} = 3 \text{ minutes}$$

$$180 \text{ seconds} \times \frac{1 \text{ minute}}{60 \text{ seconds}} = 3 \text{ minutes}$$

When the seconds aren't evenly divisible by 60, you have leftover seconds. Perform the same division and then add the extra seconds into the answer. If you want to convert 195 seconds, use the following equation:

$$195 \text{ seconds} \div 60 \text{ seconds/minute} = 3 \text{ minutes and 15 leftover seconds}$$

$$195 \text{ seconds} \times \frac{1 \text{ minute}}{60 \text{ seconds}} = 3 \text{ minutes and 15 leftover seconds}$$

The answer is 3 minutes and 15 seconds, or 3:15.

Changing hours to minutes and back again

The rules for converting hours to minutes and minutes to hours are practically identical to the minutes-to-seconds and seconds-to-minutes conversions in the preceding section.

If you have hours and no minutes, just multiply by 60 (because there are 60 minutes in an hour). For example, convert 4 hours with this equation:

$$4 \text{ hours} \times \frac{60 \text{ minutes}}{1 \text{ hour}} = 240 \text{ minutes}$$

When you have hours *and* minutes to convert, multiply the hours by 60 and add the extra minutes. A sample equation (using 2:30) is

$$2:30 = 2 \text{ hours and } 30 \text{ minutes}$$

$$2 \text{ hours} \times \frac{60 \text{ minutes}}{1 \text{ hour}} = 120 \text{ minutes}$$

$$120 \text{ minutes} + 30 \text{ minutes} = 150 \text{ minutes}$$

To do the opposite conversion, divide the number of minutes by 60 to get the number of hours. If you have minutes that are evenly divisible by 60 (using an example of 300 minutes), you can use a simple equation:

$$300 \text{ minutes} \div 60 \text{ minutes/hour} = 5 \text{ hours}$$

$$300 \text{ minutes} \times \frac{1 \text{ hour}}{60 \text{ minutes}} = 5 \text{ hours}$$

When the minutes aren't evenly divisible by 60, you have leftover minutes. Add these extra minutes into the answer. A sample equation (using 76 minutes) is

$$76 \text{ minutes} \div 60 \text{ minutes/hour} = 1 \text{ hour with } 16 \text{ leftover minutes}$$

$$76 \text{ minutes} \times \frac{1 \text{ hour}}{60 \text{ minutes}} = 1 \text{ hour with } 16 \text{ leftover minutes}$$

The answer is 1 hour and 16 minutes, or 1:16.

Working with time as a fraction

As we note earlier in the chapter, time typically appears expressed as hh:mm, mm:ss, or hh:mm:ss. But sometimes you see time expressed as a fraction, whether that's a decimal fraction, a common fraction, or maybe even a percentage. Here are some examples:

15.5 hours $3\frac{1}{2}$ minutes   50 percent of a one-hour presentation

The numbers used in denoting time and doing math about time are just that — numbers. They put on their pants one leg at a time, just like any other numbers. (Well, they would if they had legs and pants.) So the techniques for converting numbers don't vary when you work with time. For example, 50 percent of an hour is just like 50 percent of a pound of flour. And 50 percent (a percentage) converts to

$$\frac{1}{2}$$

(a common fraction) or to 0.5 (a decimal fraction).

You use the arithmetic you know to convert minutes (which are fractions of hours) and seconds (which are fractions of minutes) into fractions. You can also use the math you know and the time conversions in this chapter to convert from fractions to hours, minutes, and seconds.

When you see a number representing minutes and seconds (1:30, for example), you think of it as "one minute and thirty seconds," but it's also "one minute and part of a minute" because those leftover seconds are some fraction of a minute. Here's the conversion of 1:30 from time notation to a common fraction:

$$1:30 = 1 \text{ minute and } 30 \text{ seconds} = 1 \text{ minute} + \frac{30}{60} \text{ minute}$$

$$1 \text{ minute} + \frac{30}{60} \text{ minute} = 1\frac{30}{60} \text{ minutes} = 1\frac{1}{2} \text{ minutes}$$

Of course,

$$\frac{1}{2}$$

is also 0.5, so 1:30 is also 1.5 minutes. You get the common fraction for practically nothing, and the decimal is free.

You can find out more than you ever thought possible about fractions by looking at Chapters 8, 9, and 10.

Time Math: Calculating Time

Time math isn't difficult. It's the same arithmetic you already know — the same arithmetic covered in other chapters throughout this book. The differences between time math and normal math are

✔ The separator is a colon (:), not a decimal point (.)

✔ You borrow and carry at a different number (60) than you do with regular math (10).

✔ You don't typically have to use fractions. You express any leftover amount in an operation as just what it is, a leftover number of seconds or minutes.

The following sections give you a whirlwind course in time arithmetic.

Addition

Time addition is handy when you sum up the total time expended for any activity. It's also good when you need to add a duration to a start time to get an end time. ("If I start at 9:00 a.m. and the job takes two hours, I should be done at 11:00 a.m.") It works with a column of values just like regular addition does.

To add two time quantities together, follow these rules:

✔ Add like units, hh:mm to hh:mm and mm:ss to mm:ss.

✔ Express all factors as hh:mm:ss to make unlike units the same, filling in missing columns with zeroes as necessary to help everything line up.

For example, to add hours and minutes (such as 2:33) to minutes and seconds (such as 45:45) write them both in the same format.

2:33 becomes 2:33:00

45:45 becomes 0:45:45

Add the terms together. If the seconds sum is more than 60, carry the excess to the next column like you would in normal addition. In this example, you end up with a term of 78 minutes:

$$
\begin{array}{r}
2:33:00 \\
+\ 0:45:45 \\
\hline
2:78:45
\end{array}
$$

The 78 minutes is 1 hour and 18 minutes, so convert 60 of those minutes to an hour and carry a 1 to the hours column (filling in the minutes and seconds columns with zeroes to keep everything lined up).

$$
\begin{array}{r}
1:00:00 \text{ the carry} \\
+\ 2:18:45 \\
\hline
3:18:45
\end{array}
$$

The answer is 3:18:45.

Subtraction

Time subtraction is useful when you want to schedule events or divide your time. It's good when you need to subtract a duration from an end time to get a start time. ("If I need the casserole to be done at 11:00 a.m. and it takes two hours to bake, I should put it in the oven at 9:00 a.m.")

To subtract two time quantities, follow these rules:

✔ Subtract like units, hh:mm from hh:mm and mm:ss from mm:ss.

✔ Express all factors as hh:mm:ss to make unlike units the same, filling in missing columns with zeroes to help everything line up.

For example, to subtract minutes and seconds (such as 45:45) from hours and minutes (such as 2:33) write them both in the same format:

2:33 becomes 2:33:00

45:45 becomes 0:45:45

As with regular subtraction, if you don't have enough seconds to subtract from, borrow from the minutes column; the same goes for borrowing hours to get enough minutes. For example:

$$\begin{array}{r} 2:33:00 \\ -\ 0:45:45 \\ \hline 2:-12:-45 \end{array}$$

You can't have –45 seconds or –12 minutes, so do some borrowing. Borrow 60 seconds (1 minute) from the minutes column and 60 minutes (1 hour) from the hours column so that you have 1 fewer hour, 59 more minutes, and 60 more seconds and try the math again:

$$\begin{array}{r} 1:92:60 \\ -\ 0:45:45 \\ \hline 1:47:15 \end{array}$$

The answer is 1:47:15.

Multiplication

Multiplying time values is helpful when you need to know the total time expended for repetitive tasks. You don't multiply one time value by another time value; typically, you multiply a time value by a number. For example, to figure the total time to do three equal tasks — say, 1 hour and 42 minutes (1:42) each— use an equation such as the following:

$$1:42 \times 3 = ??:??$$

Multiply each column of units (in this case, the hours and minutes) by the multiplier (3). Be prepared to carry.

$$1:42 \times 3 = 3:126$$

Wait! You can't have 126 minutes, so convert them to hours. 126 minutes is 2 hours and 6 minutes. Add these 2 hours to the hours column and rewrite the answer:

$$3:126 = 5:06$$

The answer is 5:06.

Division

Dividing time values is useful when you want to economize on time or find new time values to fit into a busy schedule. You don't usually divide one time value by another time value, although it's possible. Typically, you divide a time value by a number. For example, say you want to cut a 1-hour-and-42-minute (1:42) task in half:

$$1:42 \div 2 = ??:??$$

Divide each column of units.

$$1:42 \div 2 = 0.5:21$$

You can't have 0.5 hours in the hours column, so convert the fractional hour to minutes. There are 60 minutes in an hour, so 0.5 hours is 30 minutes. Add the converted minutes to the minutes from your original answer and rewrite the answer:

$$0.5:21 = 00:51$$

The answer is 00:51, or 51 minutes.

Example: The Timesheet for All Trades

The timesheet is universal because completing a timesheet for payroll is common in most fields. A similar situation is when you record and accumulate time to bill to a client.

This week has been tough for you at work. Monday was okay, but your child got sick and you came in late on Tuesday. You took a long lunch to take her to the doctor (amazingly, the pediatrician was able to squeeze her in). On Wednesday, her illness continued and you couldn't get in until 1:00 p.m., so you only worked in the afternoon. The good news is that Thursday was normal and Friday was a paid holiday.

Your company doesn't believe in family leave. The company's enlightened policy is, "You're lucky to have a job, so shut up." Therefore, you get paid this week only for the hours you worked.

Figure 18-5 shows your completed timecard for this week. Calculate the daily hours you worked from the in and out times given and then add the hours to get your total regular hours. Combine those hours with holiday hours to get your total weekly hours.

PAYROLL TIMECARD

	MON	TUE	WED	THU	FRI	TOTALS
IN	8:00	9:00		8:00		—
OUT	12:00	12:00		12:00		—
IN	1:00	2:00	1:00	1:00		—
OUT	5:00	5:00	5:00	5:00		—
Regular hours	8.0	?.?	?.?	?.?		?.?
Overtime hours						
Vacation hours						
Holiday hours					8.0	?.?
TOTAL	8.0	?.?	?.?	?.?	8.0	?.?

Figure 18-5:
Weekly
timecard.

The daily sums for Monday and Friday are already filled in. To complete the timecard, do the following:

1. **Figure out your regular hours for Tuesday.**

   Subtract 9:00 from 12:00 to get 3 hours for the morning time. Subtract 2:00 from 5:00 to get 3 hours for the afternoon time. Add them together and write 6.0 in the regular hours field for Tuesday and in the total line.

2. **Determine your regular hours for Wednesday.**

   You were out all morning, so no hours there. Subtract 1:00 from 5:00 to get 4 hours for the afternoon time and write 4.0 in the regular hours field for Wednesday and in the total line.

3. **Figure out your regular hours for Thursday.**

   A nice normal day. The times are the same as for Monday, so just copy those figures.

4. **Calculate your total regular hours for the week.**

   Add the regular hours from each day and write the total (26.0 hours) in the appropriate cell in the TOTALS column.

5. **Making sure to account for your Friday holiday hours, determine your total paid hours for the week.**

   Write your holiday hours (8.0 hours) in the appropriate cell in the TOTALS column. Add the total from Step 4 to the 8 holiday hours from Friday. The sum is 34.0 hours, so. write that in the TOTAL box in the TOTALS column.

The answer is 34.0 hours. You lost 6 hours this week. It's a shame, but at least your child is getting better.

Example: Microwave Magic

This story problem is a quickie that's true for the company break room and your kitchen at home.

Today's lunch is one of those delightful little frozen sandwich pockets. You remove the plastic wrap and insert the taste-tempting treat into its browning sleeve. The instructions say to microwave for one and a half minutes, but you can't very well enter the fraction on the microwave's digital keypad. How do you get the right nuking time on the microwave?

1. **Convert one-and-a-half minutes to either time notation (mm:ss) or seconds.**

    • To convert to time notation, just follow this process:

    $1\frac{1}{2}$ minutes $= 1$ minute $+ \frac{1}{2}$ minute

    1 minute $= 00:60 = 1:00$

    $\frac{1}{2}$ minute $= 00:30$

    $1:00 + 00:30 = 1:30$

    The answer is 1:30.

    • Follow this process to convert to seconds:

    $1\frac{1}{2}$ minutes $= 1$ minute $+ \frac{1}{2}$ minute

    1 minute $\times 60$ seconds/minute $= 60$ seconds

    $\frac{1}{2}$ minute $\times 60$ seconds/minute $= 30$ seconds

    60 seconds $+ 30$ seconds $= 90$ seconds

    The answer is 90 seconds.

2. **Enter either *1:30* or *90* on the microwave oven's keypad.**

    Soon your gourmet lunch will be ready to enjoy!

Chapter 19

Math for Computer Techs and Users

*T*he personal computer (PC) has shaken the world. There were 1 *billion* PCs in the world in 2007, and *Business Week* reports that Bill Gates wants to see that figure double by 2015. You don't find as many PCs as you do cell-phones, but PCs (with their friend, the Internet) have changed business and science completely. You may even have grown up in a world that has never *not* had PCs.

Although computers often get a bad rap for being nerdy when their cousins the sports car and smartphone are sexy and cool, the computer is important in your work. Few careers can operate without them; even the artist uses a computer to run his or her art gallery, and the graphic artist today is a *computer artist*. Not only is virtually every technical career tied in some way to the computer, but numerous technical and community colleges offer special two-year degrees in Computer Technology.

Assuming that you're going to have to live with computers, you need to understand a few things about a computer's capacity and speed, so this chapter shows you some basic computer math.

You do a quick overview of numbers and units used with computing, use math to understand the computer's parts, and prepare yourself to make smart decisions when you buy a computer.

Try a Bit of This Byte: Understanding Basic Computer Terms

No matter whether you make computers your full-time career or just use them in your work, you're more computer savvy when you know a few words. Take a minute to check out the two important words always associated with computers: bit and byte.

A *bit,* also known as a *binary digit,* is the smallest unit of information a computer can store. It has only two states, off and on, which are represented by two different current or voltage levels (known as high and low).

The two states represent 0 and 1, the only numbers in the digit-challenged binary number system (which we cover in Chapter 11). But despite its lack of numerical variety, the binary system can easily manage, manipulate, and display big numbers — and all the characters of the alphabet — in a computer.

Here's an example of using binary numbers to communicate information. Yes, folks, you *can* try this at home.

Your business has a neon "Open" sign that has two states, ON and OFF. The following table shows what the state of the switch can communicate to people who drive by.

Switch	Meaning
ON	We're open.
OFF	We're closed.

Two bits can say a lot more than one bit, because they offer four possible combinations of the states (OFF/OFF, OFF/ON, ON/OFF, and ON/ON). Say you want to visit a friend's home, where the living room light and the bedroom light are each controlled by a switch. The following table shows what the states of the two switches can communicate to you when you drive by the house.

Switch 1 (Living Room)	Switch 2 (Bedroom)	Meaning
ON	ON	I'm home and I'm getting dressed.
ON	OFF	I'm home. Do drop in.
OFF	ON	It's late. I'm reading in bed.
OFF	OFF	I'm asleep. Don't bother knocking.

From analog to digital in the 21st century

Today is an age of incredible advancement in technology. Almost every day, you see or read about innovations in healthcare, pharmaceuticals, green building materials, and electronics.

You see digital replacements for analog (read: old-style) devices everywhere, including the chef's digital thermometer and digital timer, the surveyor's digital theodolite, and the automobile tech's diagnostic computer. Even the x-ray machine in the dental office displays the "x-ray" on a monitor. You can also get digital micrometers, digital body mass index measurement tools, and digital shipping scales.

You can also see (and maybe be overwhelmed by) the many general-purpose devices available. They include smartphones, digital cameras, camcorders, MP3 players, and high definition televisions. You use them both in business and at home.

The runner-up in the digital advancement game is the mobile phone (formerly called the cellphone). Mobiles are everywhere: Police use them in addition to their radios, and fishermen use them to land their boats at the port where they can get the best prices for their catches. The CIA's 2009 Factbook lists the number of mobile cellular telephone subscribers in 222 countries at nearly 4 billion! Even in a world with a population of almost 7 billion, that's a whole lotta cellphones. But for first place, the gold medal, the Heisman trophy, and the Hawaiian vacation with the matched luggage, the winner is *the personal computer*.

Just like your friend's two lights can give more complete information than your business's one sign, you can communicate more ideas when you use bits in combination. The combinations of bits a computer uses to represent numbers and characters are grouped into eight-bit *bytes*. The byte is the most common unit you use to measure a computer's capacity.

An *A* is stored on the computer as the eight bits 01000001. The number *9* is stored on the computer as the eight bits 00111001. Of course, bits and bytes get a lot more complicated than this example, but leave that part to the specialists.

People can't agree on where the word *byte* came from. One source says "The word *byte* is short for *binary digit eight*." Another source says IBM declared that it stood for "Binary Yoked Transfer Element (BYTE)." The most credible source says, "The term was coined by Werner Buchholz [at IBM] in 1956 . . . It was a mutation of the word 'bite' intended to avoid confusion with 'bit.'"

Because computers hold so many bits and bytes, the amounts of these units are often expressed with the old standby Greek prefixes we discuss in Chapter 11. You need to know kilo-, mega-, giga-, and tera- to talk computer talk. The following handy table gives you a crib sheet.

Term and Abbreviation	Basic meaning	Number of bytes
Kilobyte (K)	About a thousand	1,024
Megabyte (MB)	About a million	1,048,576
Gigabyte (GB)	About a billion	1,073,741,824
Terabyte (TB)	About a trillion	1,099,511,627,776

The Sum of the (Computer) Parts, and the Numbers Involved

To find personal happiness, you need to talk to a man in a saffron robe, sitting on top of a mountain. To find PC happiness, the answer is simpler: More is better.

With PCs, greater capacity and greater speed seem to make the difference in performance. If your PC is running slowly, you don't have enough of something. Knowing a little about the following items helps you become a well-informed computer user and computer purchaser. That's a big step toward achieving PC happiness. Om!

- ✔ Disk capacity in megabytes, gigabytes, and terabytes
- ✔ Flash memory capacity in megabytes and gigabytes
- ✔ RAM (random access memory) in gigabytes
- ✔ Processor rate in hertz
- ✔ DVD write speed
- ✔ Network speed in kilobits, megabits, and gigabits per second

Some people say they don't know anything about computers, or that they don't understand them. Then (and they're rationalizing) they say that they don't care. Don't you become part of "some people!"

Remember, the automobile and healthcare industries require bigger vocabularies and require more complicated choices than understanding computers does. You can get a handle on computers if you put your mind to it.

Disk capacity

A "good" PC has lots of storage capacity. "Lots" varies depending on whether you store mostly documents (relatively small files), mostly photos (relatively large files), or other stuff (a mix). But you can generally follow this rule of thumb: Get the biggest disk drive you can afford. This guideline applies equally to medical offices, construction company offices, and food service offices, and to laptop computers used in all trades.

Bigger is better. And disk storage is pretty cheap. Currently, most disk drives are in the 200- to 300-GB range, but offerings of 0.5TB, 1.0TB, and 1.5TB are starting to appear on the Internet and in retail stores. In extreme cases, you can put multiple disk drives ("multiple spindles") inside the case of a desktop computer or even add external storage by connecting a portable disk drive to a USB port.

Figure 19-1 shows the properties of a typical disk drive on a PC.

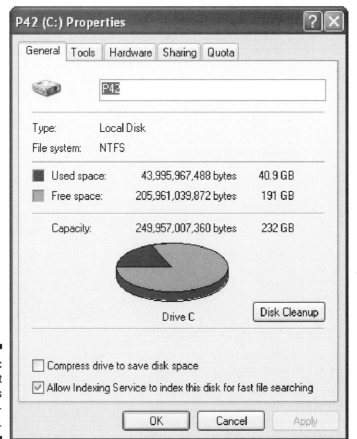

Figure 19-1:
Microsoft
Windows
disk proper-
ties display.

The math concepts you need to evaluate a disk drive are a knowledge of the units (gigabytes and terabytes) and a grasp of whole numbers. Basic arithmetic and percentages are useful, too, if you're calculating how much of the drive you're using. (You don't *have* to do calculations, but they can come in handy.)

Notice that the capacity in Figure 19-1 is 249,975,700,360 bytes. If you check your receipt or your packing slip, this figure makes sense — you probably bought a computer with a 250GB disk drive.

So why does the window also say that 232GB is a capacity? Because that's what you have left after formatting the drive. The computer requires a bit of "overhead" space for disk formatting, writing file system data (for file name, creation date, access permissions, and contents), and so on.

Figure 19-1 also shows the amount of space you've used and the amount of free space. If you want, you can calculate percent utilization. The formula is

$$\frac{\text{bytes used}}{\text{total bytes available}} \times 100 = \text{percent used}$$

To make things simple, use the number of gigabytes displayed as being available (in this case, 232), not the full number of bytes the computer started out with (250).

$$\frac{40.9}{232} \times 100 = 0.17629 \times 100 = 17.629 \text{ percent}$$

The disk is 17.6 percent full.

Flash memory

Flash memory isn't actually memory but rather computer storage on a chip. It was invented by Dr. Fujio Masuoka in 1984 when he was working for Toshiba. These days, you typically call flash memory a *flash drive*, which is a computer storage device that acts like a disk. It has a drive letter, and you manage it the same way you would any other disk on your computer. A USB (universal serial bus) flash drive gives you portable, compact storage. Flash drives are also embedded in digital cameras, handheld computers, mobile phones, music players, digital video cameras, video game consoles, and GPS devices.

You measure flash drive capacity in megabytes and gigabytes. When you buy flash drives, the math required is simply knowledge of units (megabytes and gigabytes) and whole numbers. The stores have flash drives with capacities of 64MB, 128MB, 256MB, 512MB, 1GB, 2GB, 4GB, 8GB, 16GB, 32GB, and 64GB. Some available flash drives go as high as 256 gigabytes.

Toshiba tried to demote Fujio Masuoka after he invented what became the most important semiconductor innovation in recent times, even though the chips went into products worth more than $3 trillion, including automobiles, computers and mobile phones. To cover their embarrassment, Toshiba said that Intel invented flash memory, but no, Intel says Toshiba did it.

Rama lama ding dong: RAM memory

Random access memory (RAM), also called *volatile memory,* is the memory your computer uses to run programs while it's on. After you turn the computer off, that "remembered" information is gone. Technically speaking, RAM is storage on integrated circuits (chips) on your computer's motherboard. It used to be measured in megabytes but is now described in gigabytes.

If you work in an area with a lot of power outages, you're in danger of losing your unsaved work when the electricity goes out. If that sounds like your situation, you may want to buy an *uninterruptible power supply* (UPS). A UPS has a battery and gives you about 20 minutes of time during a blackout to save your work and shut the computer down. If you don't want to buy a UPS, remember to save your work frequently.

It's good to have a lot of RAM, and it's a better idea to have a humongous amount of RAM. Microsoft Windows Vista and Windows 7 want about 1GB of RAM just to run, so your computer should have a minimum of 2GB of RAM. Three or four GB of RAM are even better.

ROM (read-only memory) isn't the same as RAM. ROM is also located in chips on the motherboard, but it holds the information used to boot up the computer.

Speed out of the gate: Processor rate

The speed of your computer's *central processing unit* (CPU) was a traditional way of determining how fast your computer was. However, times have changed. Overall computer speed is a combination of factors that we don't get into here, plus the fact that speed is purely subjective. If the computer seems slow to you, it *is* slow.

You measure processor speed in hertz (Hz), the international unit for frequency (not the rental car company). One Hz is equal to one cycle per second, which is nothing to write home about. Computers are considerably faster.

Although processor function is measured as a speed, it's not really a speed. A computer's CPU has a *clock rate,* or the number of processing cycles the CPU can perform in one second. The faster the clock, the more instructions the computer can process in that second. Today's small computers run at rates as high as 2.66 GHz. The math you need to evaluate a CPU consists of understanding the units and decimal fractions (for example 2.20 GHz, 2.40 GHz, and so on).

In the 1990s, the "speed wars" between manufacturers resulted in consumer confusion. One manufacture advertised having the highest clock rate, but the comparison was meaningless, since the competitors used a superior "architecture."

The best way to evaluate processor speed when you're buying a computer is to see whether it has one (single), two (dual), or four (quad) *processor cores*, which are separate CPUs integrated into one integrated circuit. Simply put, a quad computer is a hot computer. When you're buying, also check out the amount of L2 Cache each system has (for example, 2MB are good, but 6MB are great).

Speed is always increasing; history shows that computer speeds *double* about every 18 to 20 months. No wonder your computer seems to be out of date every time you turn around.

The Internet is running on "slow" today: Network speed

As with other components of your "computing experience" (as the marketing departments call it), network speed is important. The term *network* describes any connection between two or more computers and can include just your computer connected to the Internet, all the computers at your office, or any other linking of computers.

As the name suggests, *network speed* is simply how fast your network operates. It's measured in bits (not bytes; see the earlier section "Try a Bit of This Byte: Understanding Basic Computer Terms" for more on the difference), and faster is better. The unit of measurement for network speed is *megabits per second* (Mbps). You also see *kilobits per second* (Kbps) and *gigabits per second* (Gbps). The math for network speed consists of understanding these terms and comparing simple numbers.

Your workplace (and maybe your home) is likely to have a *local area network* (LAN). That's what connects the computers to each other (using a switch) and to the Internet (using a router and a DSL or cable modem). DSL *(digital subscriber line)* is a popular method used for sending data over a telephone line. You probably have DSL at work and at home. Cable modems hook you up to the Internet through your cable TV connection. Cable tends to be faster, but surveys indicate that customer satisfaction with cable is lower than with DSL.

Your PC can talk to other PCs and servers no faster than the speed of its internal *network interface card* (NIC). And your network switch poses an additional speed limitation. In the last few years, network speeds have gone from 10 Mbps to 100 Mbps. Now 1 gigabit per second (1 Gbps) is common. To be really fast, do the following:

✔ Upgrade your PC to a 1 Gbps NIC.

✔ Upgrade your switch to a 1 Gbps switch.

✔ When you shop for Ethernet cables, buy CAT6 or CAT5e cables. Those types are what you need for gigabit Ethernet.

✔ If you do wireless, you own a WAP (a *wireless access point*, also called a *hotspot*). Even at a "nominal" 100 Mbps, real performance tends to be 54 Mbps. Expect the same sort of diminished performance if you buy a gigabit hotspot.

The speed of your connection to other computers on the Internet (mail servers and web servers) will vary, depending on how much you pay.

You may buy DSL with varying download and upload rates. The *download rate* is how fast data is (theoretically) coming in to your computer, while the *upload rate* is how fast data is (theoretically) going from your machine to a server somewhere else (for example, your bank, eBay, or Amazon.com). Actual speed is a little lower. The following table shows you four rates from a major Internet service provider.

Plan	Cost	Download Rate	Upload Rate
Basic	$30.00	768 Kbps	384 Kbps
Express	$35.00	1.5 Mbps	384 Kbps
Pro	$40.00	3.0 Mbps	512 Kbps
Elite	$55.00	6.0 Mbps	768 Kbps

From the table, you see that you get increased download and upload speed by paying more money per month.

How fast is your DSL connection? You can find out with any of the DSL speed tests on the Internet. Try Speakeasy at `www.speakeasy.net/speedtest/`

ISDN (integrated services digital network) is another communication standard, but it has been overtaken by DSL. ISDN is costly and slow, and it's a very 1990s technology. It has some value for videoconferencing, but it has never really been popular in the United States and Canada. In computer slang, ISDN stands for "it still does nothing."

Burn, baby, burn: DVD write speed

You won't live forever, so it's nice to burn CDs and DVDs as fast as possible. Today's DVD drives allow you to use the same drive to record on both media.

With a DVD, the sign X is the unit for DVD drive speed. The math you must apply here is easy. Simply look for the X number to get a feeling for relative drive recording speed. An 8X drive is eight times faster than a 1X drive; 1X represents a recording transfer rate of 1,385,000 in bytes per second, and an 8X drives writes at 11,080,000 bytes per second. That's a lot faster. 8X is what you commonly see in stores these days, but speed figures can go up to 24X.

The rules of thumb for buying a DVD drive are simple:

- ✔ When you buy a new computer, get the DVD drive that comes with the system — that's probably an 8X DVD burner.

- ✔ When you want to record and play back high-definition video, buy a Blu-ray DVD drive. You can't play or record Blu-ray DVDs on a conventional DVD drive.

- ✔ When you're buying an external DVD drive with a USB connection (sometimes called an "outboard drive"), buy just about anything that's on sale. You determine what gives you the most value for the money.

The capacities of DVDs and CDs vary. You can find out a lot by going to the Web sites of computer manufacturers. Some (Dell, for example) show you the number of documents, presentations, photos, and MP3 files you can expect to store on each type of disk. For example, a 700MB CD can store about 175 four-minute MP3 tunes, while a 50GB Blu-ray DVD can store about 12,494 tunes. Whew! That last figure is whole lot of songs.

Older car CD players won't play MP3 files. They play AU files. Newer CD players do play MP3, but they won't play iTunes MP4 files. However, you can find free file-conversion software on the Internet.

Example: Total Capacity of a Mass Storage System

This example problem isn't for the faint-hearted, or at least that's the way it looks at first. In reality, it's not hard at all. "Difficult" calculations are often made up of smaller, easy calculations. Here, technical math meets reality:

You're a storage specialist at a medium-sized corporation and will soon take delivery of a new Hewlett-Packard XP20000 *disk array*. A disk array is a set of disk drives, ranging from 4 to 1,024 drives in number, installed in from 1 to 7 cabinets.

The array has a maximum capacity of 236 drives, but your company bought 68 drives to start. The purchase includes

- 34 fast drives (600GB drives spinning at 15,000 RPM)
- 34 slow drives (2TB drives spinning at 7,200 RPM)

As you'd expect, the salesman sold the nominal storage capacity (80.4TB), and if you multiply the number of drives by their advertised capacity, that's the answer you get. The question is, what's the *real* storage capacity of the array?

Allow for the following factors in your calculation:

- You lose 9 percent of the capacity to disk formatting (that's 0.09, leaving 0.91).
- The data protection and recovery structure causes you to lose 25 percent of the capacity (the system will write the data for every three 600GB disks on four 600GB disks and for every three 2TB disks on four 2TB disks).
- You hold out two unused hot spares (for automatic recovery) at 600GB each (one for every sixteen 600GB drives).
- You also hold out two unused hot spares (for automatic recovery) at 2TB each (one for every sixteen 2TB drives).

Coauthor Barry is originally from multicultural Los Angeles, so about this problem he'd say "¡Ay, caramba!" and "Oy vay iz mir!" in the same breath. And yet you can use any of several orderly approaches to handling the simple math required for this problem. Here's one version.

1. **Note what you already know.**

   You have the nominal capacity of the disk drives purchased (80.4TB). Set that number aside for later, just to compare with the calculated result. You also know the number of number of drives (34 2TB drives and 34 600MB drives).

2. **Remove the hot spares from your calculations.**

   They don't come into play until one of the in-use drives fails.

   - 34 2TB drives – 2 2TB drives = 32 2TB drives

   - 34 600GB drives – 2 600GB drives = 32 600GB drives

3. **Calculate the capacity of the drives after formatting (91 percent of nominal capacity).**

   You need a capacity number for each type of drive.

   - 2TB × 0.91 = 1.82TB

   - 600MB × 0.91 = 546GB

4. **Determine how many drives you lose to the data protection and recovery structure.**

   You lose 25 percent of your capacity, so 0.25×32 drives = 8 drives lost for each drive type. You now have 24 2TB drives and 24 600GB drives available for storage.

5. **Multiply the number of drives remaining by their respective formatted capacities from Step 3.**

   The formatted capacities are 1.82TB and 546GB, respectively.

   - 24 × 1.82TB = 45.68TB

   - 24 × 546GB = 13,104GB = 13.10TB

6. **Add the two capacities.**

   45.68TB + 13.10TB = 58.78TB. Your total storage capacity is 58.78TB.

Part V

The Part of Tens

In this part . . .

There are at least ten good reasons to have a Part of Tens in a *For Dummies* book. The chapters in this part are compact and filled with handy information. As the old TV commercial said, they are fast, fair, and friendly.

Take a look at Chapter 20 for a compact summary of ten problem-solving techniques. They work every time! Chapter 21 has ten common formulas and even variants and a few rough estimates for when you're in a hurry. Chapter 22 is a self-help book in itself. It has ten techniques for getting better at math while you do the other things in your daily routine.

Chapter 20

Ten Tips for Solving
Any Math Problem

*M*ath problems may look different from each other, but they're surprisingly similar. Many of the math questions you encounter in a particular career come up over and over again. For example, a graphic artist doing book design calculates cover dimensions, spine width, and margins many times for different books. Also, many similar math questions come up in all careers, especially questions about time and money.

Regardless of your field, you have to solve the math questions you come across, and this chapter gives you the principles to do just that. You may not use all of the principles all the time, but you'll use a good portion of them. Think of Sherlock Holmes, the greatest detective of all time. He investigated many different kinds of mysteries, but he consistently applied the same excellent reasoning.

Figure Out Exactly What the Problem Asks For

Know what you're trying to find. Often, the instructions clearly state what the problem is asking you to find, but sometimes the needed answer is a little obscure. People are people. They aren't perfect, and their communication with you isn't always perfect either. Here are a few guidelines for determining what the problem is asking for:

- ✔ **If you get an e-mail or a memo, read it carefully — twice — to make sure you've got a handle on all the facts it includes.** Of course, a note filled with facts and an exact question is much more helpful than one with a vague question like "How much will it cost to paint the Jones residence?", but you have to carefully examine the facts to figure out just how much info they give you.

- ✔ **If someone gives you the problem orally, carefully write it down.** In the building trades, this step is very important when you're dealing with client specifications. Your goal is to provide exactly what the client wants, and that's a lot easier to remember when you have it in writing.

- ✔ **Find the keywords.** Look for phrases like "How much?", "When?", and "How many?" They clue you in to what the problem wants.

- ✔ **Start to form a mental picture of how you may find the answer.** Do you need to count? Calculate areas? Use simple arithmetic? Use a trigonometry triangle? For example, if the question is about the cost of carpeting five rooms, it suggests that you must calculate the area to be covered before figuring a price.

List the Facts

Identify and list the facts. If you go over all the information, you find (or should find) most of what you need. From this list, you convert the information, find what's missing, and eliminate what's excessive. We cover these steps in the following three sections.

Get organized and stay organized. You don't want to overlook any facts, and messy note-taking may cause you to lose facts. If you're doing the work on paper, tiny sticky notes don't cut it. Get a nice big legal pad.

Convert Supplied Information into Needed Information

Sometimes information is like a bad Christmas gift — it comes in the wrong size, shape, or color. You may have plenty of information to solve a problem, but it isn't worth anything until you make it usable. Luckily, you may be able to convert the facts you have into the facts you need.

For example, in a medical office, you may convert from metric to American units to describe how much of a substance a patient should take. The Institute of Medicine (the health arm of the National Academy of Sciences) recommends that pregnant women drink 2.3 liters of water per day. But most Americans measure in ounces, so you need to be able to convert that figure to let your patients know how many 8 ounce glasses that equals. (The answer is 9.7 glasses per day.)

Sometimes you have to convert subjective data into hard facts. For example, if a painting client says, "I want to do the bedroom in sort of a robin's egg blue," you need to convert that information into a fact — the exact paint color and brand — before you can price the job.

Determine What Information You're Missing

The fact list and conversions from the preceding two sections help you see the information you have. What remains to find is the information you're still missing. Your mind will say, "Hey, wait a minute!" as you realize that the problem can't be solved without more information.

Sometimes filling in this info is a matter of (diplomatically) asking the boss or the client, "Hey, what did you mean by this?" It's better to ask questions than to make a major error because of missing facts. In the worst case, you can do two things: assume or estimate. Sometimes these techniques are necessary, but they're never as good as having the facts.

Eliminate Excess Information

Sometimes you get all the information you need and then some. When you get a lot of information, dismiss any excess information that doesn't help you find the solution.

When people express a math problem verbally, be an active listener. When a concern comes up in a staff meeting, the real problem may be hiding in the fluff — opinions, discussions about personalities, or office trivia. Filter out the excess, but stay open to minor facts that affect your solution. You can do the same thing with written problems, too.

For example, you may think a statement like "Joe has the exact prices for the fencing lumber, but he's on vacation in Hawaii" has nothing to do with your problem, but it's actually a signal to you that you can't rely on Joe for costing a job. Instead, you need stored information (maybe your own experience, previous contracts in paper files, or spreadsheets stored on your computer), online sources, and maybe an Internet calculator. No, Joe's being in Hawaii doesn't actually help you figure out the lumber cost, but it does let you know that he isn't available as an info source for your problem-solving process.

Draw a Diagram

Use a diagram to help visualize the problem. The expression "Let me show you" is very powerful, and you often see things in a simple drawing that you'd miss just looking at numbers.

A diagram is often useful for visualizing counting problems, including inventory, parts distribution, and patient dosing. Just draw some circles on a piece of paper. The circles represent locations. Use symbols that place the items in different locations. This method can work for lubricants in multiple auto service locations, tool inventory for multiple CNC work modules, emergency medical supply pre-need placements, pantry stocking at multiple kitchens, and so forth.

Use the charts described in Chapter 17 if they apply. You may also want to make a table of values. Draw floor plans or property plans if you need to. When you sense that you must calculate the relative proportions of "things," make a simple pie chart. A pie chart shows the proportion of things you have to each other.

Find or Develop a Formula

When you see the nature of the problem, find or develop a formula to solve it. (See Chapter 13 for the discussion of the time/distance/speed standard formula or the burgers-and-fries custom formula.) Every problem requiring math has an underlying formula. Even counting three apples uses the formula $x = 1 + 1 + 1$.

Some of the calculations you do are so commonplace to you that you probably have memorized them. For example, a cement truck holds eight cubic yards of concrete. You can use that formula (1 truck = 8 CY) every time when you're planning a pour.

Some calculations rely on a well-established formula. For example, there's a formula for the feed rate of a machine tool (Inches per minute = RPM × Inches per flute × Number of flutes). If you know your material and your tool bit, you can customize the formula and keep it forever. For example, if you're cutting 0.19 inches deep into 3003 aluminum, using a .25-inch two-flute bit at 20,000 RPM, you can go at about 80 inches per minute.

Don't forget real life experience. For example, if the feed rate breaks the tool (which it shouldn't) or damages the workpiece (which it shouldn't), you've learned a valuable lesson and know to modify the formula next time.

After you have formulas for the problems you regularly solve, calculation is fast and easy.

Consult a Reference

No one has a perfect memory, and no one expects you to have one either. Rely on your external memory. That is, consult references as needed. For example, only a few older machinists have memorized the decimal equivalents of the 63 fractional drill sizes from 1/64 inch to 63/64 inch. This memorization used to be commonplace, but today most machinists use a conversion chart.

If your company has printed reference materials (such as lumber price guides, metal machining specifications, or successful recipes), take advantage of them. You can also go to the Internet to find many conversion charts, conversion calculators, and facts about materials. Just remember that the Internet is far from perfect. Do some research before you place your trust in a single Web site.

Pay Attention to Units

Watch units and unit conversions. They can throw an answer way off. Combine like units and don't combine unlike units; feet and meters (each from a different measurement system) don't combine. For that matter, feet and miles (both from the American measurement system) don't combine, either. Because you need like units to do the math, be prepared to convert units to get compatible figures.

When you have a final answer, be prepared to convert units to a form that clients and vendors can understand. For example, a recipe may call for 384 fluid ounces of cream, but your supplier expects you to order 3 gallons. You can use a table of conversions on the Internet, a calculator, or your own memory to find conversion factors.

Check Your Answer to See whether It's Reasonable

Test your answers for reasonableness. Experience is a great help, so if you don't have a lot of experience, consult with a co-worker. For example, if you calculate concrete needed to pour a residential patio and the answer is 14 truckloads, you probably messed up. That's 112 yards! The units are wrong or the math is wrong; maybe you didn't convert the cubic feet you got calculating the patio's volume into cubic yards, the number that the concrete company needs.

You can recalculate the math used in a problem in numerous ways, including adding columns from the bottom up or turning subtraction results into an addition problem. Test your results. And by keeping your work neat, you more easily see incorrect units or a bad unit conversion.

Chapter 21

Ten Formulas You'll Use Most Often

· ·

In This Chapter

▶ Identifying popular math formulas

▶ Using common conversion factors

· ·

*Y*ou can find thousands of math formulas. Higher mathematics has many branches, including the study of numbers, sets, probability, and statistics, and they all have formulas. The good news is that technical math uses far fewer formulas, and only a few of those come up all the time. This chapter lists the ten formulas you probably use most frequently and shows you some shortcuts. As a bonus, you often can use one formula two or three ways.

Area of a Square, Rectangle, or Triangle

In some trades, you find areas all the time, so at some point, you invariably need to calculate the area of a rectangle, square, or triangle (such as to pave a parking lot or landscape a yard). The most common unit of square measure is the square foot. (Chapter 15 gives you more information on areas.)

✔ **Area of a rectangle/square:** To calculate the area of a rectangle or a square, the formula is $A = L \times W$, where L is the length and W is the width.

A handy shortcut: When you know two of the terms, you can use this formula to find the third term. If you know the area and the length, you can calculate the width. If you know the area and the width, you can calculate the length.

✔ **Area of a triangle.** To calculate the area of a triangle, the formula is

$$A = \frac{1}{2}(b \times h)$$

where b is the base and h is the height.

As a shortcut, when you know two of the terms, you can use this formula to find the third term. If you know the area and the base, you can calculate the height. If you know the area and the height, you can calculate the base.

Area of a Circle

Many trades (such as machining and landscaping) work with surfaces shaped in circles, so knowing how to find a circle's area comes in handy. Flip to Chapter 15 for more on dealing with the areas of circles.

To calculate the area of a circle, the formula is $A = \pi r^2$ ($A = \pi \times r \times r$), where r is the radius of the circle and π is 3.14159.

If you know the diameter of the circle but not the radius, remember that the radius is one-half the diameter.

If you want a rough shortcut calculation, use

$$3\frac{1}{7}$$

for pi.

A handy shortcut: When you know the area, you can use this formula to find the radius.

Feet to Meters and Inches to Centimeters

Metric units and American units don't mix (see Chapter 6 for more on these systems), so the ability to convert measurements from one system to another is handy. The two most common small measurement conversions are feet to meters and inches to centimeters.

✔ **Feet to meters:** The conversion factor for feet to meters is 1 foot = 0.3048 meters.

All you have to do is multiply your measurement in feet by 0.3048 to get the measurement meters. 100 feet, for example is 30.48 meters. A good estimation shortcut is that a foot is about one third of a meter.

✔ **Meters to feet:** The conversion factor for meters to feet is 1 meter = 3.28 feet. Multiply your measurement in meters by 3.28 to get the measurement in feet. A 100-meter race is 328 feet long. As a rule of thumb, a meter is about 39 inches.

✔ **Inches to centimeters.** The conversion factor for inches to centimeters is 1 inch = 2.54 centimeters. Just multiply your measurement in inches by 2.54. You can also estimate that four inches equals about ten centimeters.

✔ **Centimeters to inches.** The conversion factor for centimeters to inches is 1 centimeter = 0.39 inches. Multiply your measurement in centimeters by 0.39. A child who is 152 centimeters tall is about 60 inches tall. As a shortcut in converting centimeters to inches, just multiply centimeters by 4 and divide the result by 10.

Miles to Kilometers and Kilometers to Miles

The difference between metric and American units can make measuring distances difficult. Automobile odometers still record miles, but for many years car speedometers have had kilometers per hour on a scale next to miles per hour.

If you work with distances (for example, in surveying and fighting wildfires), or drive as part of your work (for example, as an emergency medical technician), converting between American and metric units may become very important.

✔ **Miles to kilometers:** The conversion factor for miles to kilometers is 1 mile = 1.609 kilometers. Multiply your miles measurement by 1.609 to get the distance in kilometers. For example, when a city is 100 miles away, it's 160.9 kilometers away.

You can make a very rough conversion of miles to kilometers by multiplying miles by 1.5.

✔ **Kilometers to miles:** The conversion factor for kilometers to miles is 1 kilometer = 0.6214 miles. Multiply your measurement in kilometers by 0.6214. As a shortcut, a kilometer is about 0.6 of a mile, so if a city is 100 kilometers away, it's about 60 miles (actually 62.14 miles) away.

Pounds to Kilograms and Ounces to Grams

If you work in the culinary field, you may run across recipes with ingredients measured in another unit system. (Check out Chapter 6 for more on measurements and conversions.) The following conversions can help you make the switch.

- **Pounds to kilograms:** The conversion factor for pounds to kilograms is 1 pound = 0.454 kilograms. All you need to do is multiply the number of pounds by 0.454. For example, 100 pounds of sugar weigh 45.4 kilograms.

- **Kilograms to pounds:** The conversion factor for kilograms to pounds is 1 kilogram = 2.2046 pounds. Multiply your measurement in kilograms by 2.2046 to get pounds. A person who weighs 100 kilograms weighs 220 pounds.

- **Ounces to grams:** The conversion factor for ounces to grams is 1 ounce = 28.35 grams. Just multiply the number of ounces by 28.35.

- **Grams to ounces:** The conversion factor for grams to ounces is 1 gram = 0.3527 ounces. Multiply your measurement in grams by 0.3527 to convert to ounces.

Gallons to Liters and Liters to Gallons

In the United States, gasoline is sold by the gallon, but it's sold by the liter in many other countries. Chefs also often find liquids labeled with both American units (quarts and fluid ounces) and metric units (liters).

- **Gallons to liters:** The conversion factor for gallons to liters is 1 gallon = 3.785 liters. Multiply the number of gallons by 3.785 to get the number of liters.

    If you have a measurement in quarts but not gallons, remember that a gallon contains four quarts. Multiply your quart measurement by 0.9463 (that's also the same as 3.785 ÷ 4) to get the liter measurement.

- **Liters to gallons:** The conversion factor for liters to gallons is 1 liter = 0.264 gallons. All you do is multiply your measurement in liters by 0.264 to get gallons.

Temperature Conversions

Temperature in the United States is traditionally measured using the Fahrenheit scale, but if your work includes contact with people in other countries (or working with instructions written by those in other countries), you are likely to encounter temperatures expressed using the Celsius scale. It's useful to be able to convert from one system to the other.

- ✔ **Fahrenheit to Celsius:** To convert a temperature in degrees Fahrenheit to a temperature in degrees Celsius, subtract 32 from the Fahrenheit temperature and then multiply the result by $\frac{5}{9}$. Here's what the formula looks like:

$$\text{degrees C} = \left(\text{degrees F} - 32\right) \times \frac{5}{9}$$

- ✔ **Celsius to Fahrenheit:** To convert from degrees Celsius to degrees Fahrenheit, just multiply the Celsius temperature by 9/5 and add 32 to the result. Here's that formula:

$$\text{degrees F} = \left(\text{degrees C} \times \frac{9}{5}\right) + 32$$

Hours to Minutes and Minutes to Hours

The need to convert hours and minutes occurs in all trades when you calculate client billing or hours worked. In the culinary arts, time conversion is especially important, as cooking often requires several different dishes to be prepared simultaneously.

- ✔ **Hours to minutes:** As you've probably heard, there are 60 minutes in an hour, so to convert from hours to minutes, just multiply the number of hours by 60 to get the number of minutes.

- ✔ **Minutes to hours:** To convert minutes to hours, simply divide the number of minutes by 60.

When the minutes aren't evenly divisible by 60, you have leftover minutes. Add these extra minutes into the answer. For example, 76 minutes converts to 1 hour, 16 minutes.

Distance, Time, and Speed

If your work includes driving or flying and it's time-sensitive, it's essential to be able to calculate distance based on time and speed. Critical applications include delivering medications to hospitals and nursing homes and transporting organs for transplant.

Time, speed, and distance are related, so when you know two of the terms, you can use one of the following formulas to find the third term.

- ✔ **Distance.** The formula for distance is $d = v \times t$, where v is the velocity and t is time. When you know velocity and time, you can calculate distance.

- ✔ **Speed.** The formula for speed is

$$v = \frac{d}{t}$$

where d is the distance and t is time. When you know distance and time, you can calculate velocity (speed).

- ✔ **Time.** The formula for time is:

$$t = \frac{d}{v}$$

where d is the distance and v is velocity. When you know distance and velocity, you can calculate time.

Volts, Amps, and Watts

Volts, amps (amperes), and watts are related terms used by electricians. In a typical office or home, if you plug in too many computers, space heaters, power tools, irons, or other appliances into outlets on the same circuit, a breaker trips as a safety precaution.

Volts, amps, and watts are related. When you know two of the three terms, you can use one of the following formulas to find the third:

- ✔ **Watts:** The formula for wattage is $w = a \times v$ (watts = amps × volts), where a is the amperage and v is the voltage.

  Multiply a circuit breaker's amps (typically 20 amps) by the power company's voltage (typically 120 volts) to get the number of watts the circuit can carry. For the circuit described here, the maximum wattage is about 2,400 watts.

✔ **Amps:** The formula for amperage is

$$a = \frac{w}{v}$$

where *w* is the wattage and *v* is the voltage.

Divide the wattage on a circuit (the sum of the wattages of all lights, appliances, heaters, tools, and so on connected to that circuit) by the voltage. The result is the number of amps the circuit is drawing. If the amount of amps drawn is greater than the circuit's rating, the breaker should trip any second now. Keep a flashlight handy.

✔ **Volts:** The formula for voltage is

$$v = \frac{w}{a}$$

where *w* is the wattage and *a* is the amperage.

The best way to measure voltage is with a voltmeter. With voltage, even the professional electrician may get an occasional surprise (but hopefully not a shock). We know of one situation where a normal 120-volt circuit was wired for 240 volts. That's dangerous — it can cause fire, injury or death.

Chapter 22

Ten Ways to Avoid Everyday Math Stress

In This Chapter

▶ Beefing up your math skills in everyday settings

▶ Practicing math without the pressure

*Y*ou may think you can't master math, but the prospects are good that you can. The keys to doing so are knowing the principles and practicing them. This chapter has 10 easy tips for getting better at math.

If you have trouble doing simple math (such as adding, subtracting, and multiplying), you may have a little anxiety. If you avoid math problems and mathematicians, you may even have a little fear of math (math phobia). See Chapter 2 for more on math phobia.

Remember: We authors aren't doctors, nor do we play them on TV, so you should seek medical help if you have a serious learning problem. But for ordinary math stress, the cure is right here. Take two of these tips and call us in the morning. Take all ten and see whether you don't do better at math. If your improvement persists for more than 30 years, please call us to let us know!

Get Help with Your Checkbook

Balancing a checkbook improves your addition and subtraction skills. At one time or another, *everyone* has trouble balancing a checkbook. If you get stuck, get someone to help you (as long as you don't care whether they know about your finances) or try these techniques:

✔ Keep your check register complete and up to date. Always bring down the balance. Subtract checks and add deposits.

✔ When your statement arrives, check off the cleared items on the statement and in the register. Draw a little circle next to uncleared items.

Draw a line in the register under the last cleared item shown on the bank statement. That's the date you need to reconcile to.

✔ On the back of the statement, write the bank's ending balance. In a column underneath it, write the uncleared items from your register. Subtract the uncleared checks from the balance and add uncleared deposits to the balance.

✔ On the back of the statement, write your register's ending balance (where you put the underline). In a column underneath it, write any charges or interest from the bank statement. Add interest. Subtract charges.

The sums of the two columns should be equal. If they're not, you're going to have to check things further. You may have failed to record a check or a deposit.

✔ After reconciling, be sure to write any new bank charges, interest or adjustments in the register.

Use Grocery Shopping to Build Confidence

Estimating groceries improves your estimating skills, rounding skills, and your capacity to do math in your head. If you have totally full cart, that can be quite difficult, but if you're buying just a few items, build your math skills with these tips:

✔ **Make a list before you shop.** Write the approximate cost of each item on the list. If you don't know the price, estimate based on your past experience. If a head of lettuce cost $1.00 last week, it shouldn't have risen to $1.50 this week (you hope).

✔ **Go to the grocery store with one bill (say a ten-dollar bill) in your pocket and try not to spend any more than that.** If you buy more than ten dollars' worth of stuff, you find out pretty quickly at the checkout. This experience will make you a better estimator.

✔ **As you put items in your cart, round the prices.** Always round up. Think of an $0.89 can of beans as costing $1.00 — rounding helps account for sales tax. Check out Chapter 9 for info on rounding.

✔ **Rely on experience with items you buy every week.** For example, if a 1.25 liter bottle of your favorite sparkling mineral water is never higher than $1.35, you can be pretty sure that two bottles will always cost less than $3.00 and four bottles cost less than $6.00.

✔ **Keep track of your estimate as your cart fills up.** Otherwise, you're likely to forget the estimated amount.

✔ **Count the items in your cart.** That's simple, and the tally also tells you whether to get in the express checkout line.

Practice Reading Analog Clocks

Sure, it's not as easy as glancing at your computer's digital clock, but reading *analog* clocks (those with hands and a 12-hour face) improves your estimating and your capacity to do math in your head. Here are a couple of ideas for getting comfortable with analog clocks:

 ✔ **Work with analog clocks to get better at reading and estimating time.** Become a clock watcher. Practice, and *count* the minute marks if necessary. Go up to the clock face and say the time and the duration you're measuring. For example, at 10:10 a.m., say "It's now 10:10. When Mickey's big hand moves ten marks, it'll be 10:20."

 ✔ **Develop a sense of when the clock shows the passage of 15 minutes (a quarter of an hour) and 30 minutes (a half hour).** Those are simple, common time periods that can help you better estimate how long you have until a deadline and so on.

The first Mickey Mouse watch appeared in 1933, and you can still buy them today. The Ingersoll-Waterbury Watch Company first produced it, selling over 11,000 the first day. It saved the company from bankruptcy.

Play Games

Playing games improves strategy, conceptual, counting, and addition skills. Do you have trouble keeping score during games or remembering how to keep score? The keys to improvement are to recognize the scoring pattern and to practice, practice, practice. Just think of scoring as mostly counting with a little addition thrown in. Check out these pointers for playing your way to better math skills:

 ✔ **Start by playing games with simple scoring.** In many games, your score goes up by one when you win a point, and your opponent's score goes up by one when he wins a point. It's simple counting. If you can't remember the score, just write it down.

 ✔ **Play computer games.** Many games keep the score for you, but they also teach problem-solving strategies. Try hidden object games, where you try to find many items camouflaged in a scene — these games help your perception, which is valuable in dissecting story problems or figuring out which formulas to apply. If you play games on the computer at work, just tell your boss that you're building your math skills (though that may or may not get you off the hook).

✔ **For a bigger challenge, practice scoring for games or sports like tennis, bowling, poker, cribbage, or bridge.** Games like these have special scores and odd patterns that can help you build more advanced skills. Bridge, for example, is almost incomprehensible, with points for each trick (depending on suit), overtricks, undertricks, slam bonus, honors, game, rubber, and whether you're vulnerable or not.

Memorize Math Signs, Symbols, and Formulas

If you think you can't remember mathematical concepts, rules, or formulas, don't feel bad. You remember the concepts you use most often — we bet you've memorized your phone number and know what a dollar sign ($), stop sign, and so on are — so focus on remembering the math you use all the time.

Remember, the signs for arithmetic are $+$, $-$, $\times$, and $\div$. They represent addition, subtraction, multiplication, and division, respectively, and come up in pretty much every career, so you want to know them to avoid confusing yourself and others on the job. On the other hand, signs like the square root sign ($\sqrt{\ }$) rarely come up in your work, so you don't need to devote as much attention and memory to them.

The same idea applies to formulas. For example, if you calculate a lot of areas, you know in your heart-of-hearts that the area of a rectangle is the product of the length and the width $(A = L \times W)$.

If other formulas come up in your work, you don't necessarily have to memorize them. Use common sense. Just write them down and be prepared to look them up in your manual notes, on your computer, or on the Internet.

Make the Multiplication Table a Mantra

You can do math faster in your head when you know the multiplication table, but lots of people have trouble with multiplication tables. In Chapter 5, we offer a 9 x 9 version of the table showing the result of multiply numbers from 1 through 9 by numbers from 1 through 9. It also includes multiplication by 0. If you have trouble multiplying, make a copy of that table and say the products few times: "2 times 3 equals 6," "2 times 4 equals 8," and so forth. This strategy is *rote learning*, and it works amazingly well.

Make the multiplication table your mantra. (A *mantra* is a word, phrase, syllable, or sound that you methodically repeat many times.) Say the table softly or say it loudly, but say it often. Reciting a multiplication table is easier than practicing violin. You can recite the table while you're jogging or doing dishes; you can't practice violin when you're doing those tasks.

Use Paper Maps and Practice Navigating

Do you have trouble differentiating between left and right? The condition seems to be growing more common as everyone gets more stressed by everyday life. If left and right are a problem for you (and even if they aren't), you may also have trouble reading a map — many people do. A map is just a paper picture of geography (the city of Denver, for example) on a plane surface. It's a simple case of geometry. A map also has a scale, a legend, symbols, and colors. Why, those very terms come up in Chapter 17, the chapter about graphs, and you can use them in map reading.

Getting from here to there on a map isn't difficult. Just remember the following:

- ✓ **Hold the map so you can read it.** For most roadmaps, that means north is at the top.

- ✓ **Find where you are and where you want to go, and then use a highlighter to mark the route.**

- ✓ **If you want, use the scale to get an idea of the mileage between various points on the route.** The sum of the mileages gives you the total mileage to your destination (and puts your adding skills to use).

- ✓ **Practice by using a map often, even to get to places you already know.**

Try to Estimate Distances

You may have trouble estimating measurements or distance. Why? Because just about everyone has trouble estimating measurements or distance. It's not easy to estimate 100 yards or even 10 feet! Try it. Take a tape measure, go outdoors, and estimate a distance. Then measure the actual distance with the tape. Your estimate probably isn't even close. It takes lots of practice to get good at estimating distances.

Even small measurements can be a challenge. Ask someone to hold her hands 1 foot apart. Then measure the distance with a ruler. Usually, the person's estimate is 2 to 4 inches off. Estimating a yard is even harder.

Use a shortcut. A *cubit* (about 18 inches) is the distance from your elbow to the extended tip of your middle finger. A yard is the distance from your nose to the extended tip of your middle finger. You can use your body as a measurement system. For more information, just perform a quick Web search for "your body ruler."

Take Up Music

Some people have trouble with *sequential processing*, the ability to register, collect, process, and use information in an organized way. This problem can affect both abstract tasks (for example, reading, writing, and math) and physical tasks such as doing dance steps.

Research suggests that people are involved with music do better at math. Dancing is a good way to get involved, and dance steps usually require execution in a pre-determined sequence. So, dancing is a great way to get a grip on sequential processing (which you use in virtually all math problems) Try salsa, line dancing, contra dancing, or square dancing. Learn the Texas Two-Step and Cotton-Eyed Joe. They're all fun.

If you're absolutely dance challenged, get a copy of the song "This Old Man" (a traditional children's marching song) and sing it:

> This old man, he played one,
> He played knick-knack on my thumb;
> Knick-knack paddywhack,
> Give a dog a bone,
> This old man came rolling home.
>
> This old man, he played two,
> He played knick-knack on my shoe;
> Knick-knack paddywhack,
> Give a dog a bone,
> This old man came rolling home.
>
> This old man, he played three,
> He played knick-knack on my knee;
> (Etc.)
>
> This old man, he played four,
> He played knick-knack on my door;
> (Etc.)

Notice that this old tune has a sequence built right in to it.

Integrate Math with Nonmath Skills

If you're in school, you probably do better in some subjects than in others. Think about the subjects you like and use them to build math skills.

Find the parts of your nonmath courses that require a little math. Then perform (don't avoid) the math. Try making anything into a little math exercise (for example, sequencing history dates or counting chapters). This strategy is a painless way to build your math skills. As Mary Poppins says, "A spoonful of sugar helps the medicine go down."

Even if they're not your favorites, look at subjects that emphasize thinking and logic; they may also offer opportunities to practice math. The more you integrate your math skills with your nonmath skills, the easier math becomes.

Glossary

*H*ere's a glossary of common terms you may come across in relation to mathematics, especially in the working world:

absolute value: The numerical value or worth of a number regardless of its sign; its distance from zero. For example, the absolute value of both +5 and –5 is 5.

acute angle: An angle that measures less than 90 degrees.

additive inverse: Number with the same numerical part but the opposite sign (plus or minus) of the given number. For example, if zero is the sum of two numbers, the two numbers are *additive inverses* of each other.

adjacent angles: Angles that share a vertex (corner) and have a side in common.

adjacent sides: Sides of a polygon that share a vertex.

altitude: Also known as the *height* of a triangle; a segment drawn from a vertex of a triangle to its opposite side so that it forms right angles with the opposite side.

angle: A figure formed by two rays that have a common endpoint.

angle bisector: A ray that cuts an angle in half.

arc: A portion of a circle's edge, or circumference.

area: Measure of a specified region in a plane.

associative property: Characteristic of addition and multiplication that allows the grouping of terms to change without affecting the result.

base: Value multiplied repeatedly in an exponential expression.

binary operation: Process requiring two values to produce a third value.

binomial: Two terms separated by addition or subtraction.

bisect: To divide a segment or an angle into two congruent (equal) parts.

central angle: An angle whose vertex is the center of a circle.

chord: Any segment that joins two points on a circle.

circle: A set of points in a plane that are all equidistant from a single point (the circle's center).

circumference: Distance around the outside of a circle.

coefficient: Number multiplied by a variable.

collinear points: Points that lie on the same line.

combinations: Method of counting that tells how many ways a designated number of objects can be selected from a given set.

common denominator: Same value on the bottom of more than one fraction.

commutative property: Characteristic of addition and multiplication that allows the order of the values in an operation to be changed without affecting the result.

complementary angles: Two angles whose measures add up to 90 degrees.

composite number: Whole number larger than one that isn't prime.

concentric circles: Circles that share the same center but aren't congruent.

conduction: The transmission of heat or electricity through a material.

cone: A solid three-dimensional figure that has a circle for its base and comes to a single point at its top.

congruent: Identical in shape and size (equal).

congruent angles: Angles that have the same measure.

congruent circles: Circles that have congruent radii.

consecutive: Items or elements that follow one another in some sequence.

constant: Variable or number that never changes in value in an expression.

contraction: A function in which distances are shortened.

constraint: A qualification or limiting rule that affects the values that a variable may have. For example, the variable *x* may be limited to numbers greater than 2.

construction: A drawing made using a compass and straightedge.

coordinate: Part of an ordered pair that designates a point's location on a coordinate plane.

coordinate (Cartesian) plane: Flat region determined by two intersecting, perpendicular, numbered lines called *axes.*

coplanar: Points or shapes lying in the same plane.

corresponding angles: When a transversal cuts across two parallel lines, non-adjacent angles on the same side of the transversal, one of which is between the parallel lines and one of which is outside the parallel lines. Corresponding angles are congruent.

cube: Third power of a number, result of multiplying a number by itself three times; a three-dimensional solid with six square surfaces.

cube root: Number that can be multiplied by itself three times to get a given number. For example, the cube root of 8 is 2 because 2 multiplied by itself three times equals 8.

cubic: An expression in which the highest power is three.

current: The measurement of electric charge flowing per second.

cylinder: A three-dimensional figure that has two congruent circular bases that are parallel.

decimal: Fraction with an unwritten denominator of 10 indicated by the decimal point; number system that is organized in increments of 10.

degree: A common unit of measurement for an angle; there are 360 degrees in a circle.

degree of an angle: Number of degrees between 1 and 360 that comprise the measure of an angle.

degree of an expression: Highest power occurring in the expression.

denominator: Bottom number of a fraction.

density: A quantity divided by a length, area, or volume, such as mass per length or mass per volume.

diagonal: A segment that goes from one vertex of a polygon to a non-adjacent vertex.

diameter: Longest distance across a circle.

difference: Result of subtraction.

digit: Numerals from 0 through 9.

dilation: A function in which distances are lengthened.

distributive property: Characteristic of multiplication and addition that allows for the multiplication of each individual term in a grouped series by a term outside of the grouping without changing the value of expression.

divisible: One number can be divided by another with no remainder.

edge: A segment where two faces of a three-dimensional shape meet.

energy: The ability of a system to do work.

equation: Mathematical statement with an equal sign showing that two values are equal.

equivalent fractions: Fractions equal to each other, even though they may have different denominators.

exponent: Also known as *power;* value in smaller type found above and to the right of the base that indicates repeated multiplication of the base.

expression: Combination of values (variables, numbers, constants) and operation(s).

exterior angle: An angle that's adjacent and supplementary to an interior angle or polygon.

face: A flat side of a three-dimensional shape; faces are polygons.

factor: One of the values in a multiplication operation; to rewrite an algebraic expression as a product.

factorial: Operation that multiplies a whole number by every counting number smaller than it.

foot: The point where a line intersects a plane.

formula: Rule or method that is accepted as true and used repeatedly in common applications.

fraction: Quantity expressed as a numerator (the value on top of the bar) and a denominator (the value below the bar that determines how many make one).

frequency: The number of cycles of a periodic occurrence per second.

geometric mean: The number between two other numbers that is the positive square root of the two numbers. For example, the geometric mean of 4 and 9 is six, because 6 is the square root of $4 \times 9 = 36$.

graph: Plotted figure in a plane.

grouping symbol: Parentheses, brackets, and braces that can affect the order of operations. Terms and operations within the grouping symbols take precedence.

homogeneous equations: A system of equations, all set equal to zero. The system always has a solution; the solution is *trivial* if all the variables are equal to zero and *nontrivial* if some of the variables are not equal to zero.

hypotenuse: Longest side of a right triangle.

inconsistent equations: A linear system that has no solution.

inequality: Relationship between two unequal values.

infinite: Without end; uncountable.

integer: A positive or negative whole number or zero; numbers starting with zero and going up or down in increments of one.

intercept: Point where a graph crosses the x-axis or y-axis.

intersection: Point shared by two lines.

inversion: A permutation in which a larger integer precedes a smaller integer.

irrational number: Number with no fractional equivalent whose decimal never repeats or terminates.

isosceles triangle: A triangle with at least two congruent sides and angles.

line: All the points in the coordinate plane that satisfy a linear equation.

line segment: A portion of a line that has finite length; the two ends are called endpoints.

linear: Adjective describing an expression or equation in which the highest power of any variable is one. Constants can be higher powers. For example, $x + y = 4$ is linear.

magnitude: The size or length associated with a vector.

mass: The property that makes matter resist being accelerated.

matrix: A rectangular array of numbers of elements with m horizontal rows and n vertical columns.

median: Segment that joins the midpoints of a trapezoid's nonparallel sides, or that connects a vertex of a triangle with the midpoint of the opposite side.

midpoint: A point that divides a segment into two congruent segments.

mixed number: Improper fraction written as a whole number alongside a fraction.

monomial: An expression with only one term.

multiple: A number evenly divisible by a specific factor. For example, the numbers 14 and 21 are multiples of 7.

natural numbers: Also called *counting numbers*; values starting with 1 and increasing by 1.

negative number: Any quantity that is less than zero; usually preceded by a minus sign (–).

negative reciprocals: Two numbers, one positive and one negative, whose product is –1.

non-coplanar: Not lying in the same plane.

null space: The set of all the solutions of a system of homogeneous equations.

numerator: The top number in a fraction.

obtuse angle: An angle that measures more than 90 degrees and less than 180 degrees.

obtuse triangle: Triangle that has one obtuse angle and two acute angles.

operation: Mathematical process (addition, subtraction, multiplication, division) performed on one or more quantities.

opposite: With operations, another operation that gets you back where you started. With numbers, the additive inverse, or the same absolute value of a number with a different sign.

ordered pair: Two values inside parentheses and separated by a comma that indicate the position of a point in the coordinate plane.

origin: Point of intersection of the x-axis and y-axis in a coordinate plane.

parallel lines: Lines that never intersect and are always the same distance apart.

parallelepiped: A polyhedron in which all of the faces are parallelograms.

parallelogram: A four-sided figure that has two pairs of parallel sides.

parameter: A variable used to express a relationship and whose value distinguishes the various cases.

pentagon: A five-sided polygon.

percent: Fractions with a denominator of 100. The percentage is the numerator of the fraction, or how many out of 100.

perimeter: Total distance around the outside of a region or area.

permutation: Counting method that determines the number of ordered arrangements there are when a certain number of objects are selected from a given set.

perpendicular lines: Lines that form a 90-degree angle at their intersection.

plane: A flat, infinitely thin surface that goes on forever in every direction.

point: An infinitely small dot.

polygon: A closed shape with straight sides and no gaps or openings.

polyhedron: A multisurfaced figure.

polynomial: Expression with one or more terms.

positive number: Any quantity greater than zero.

postulate: A geometrical statement that's assumed to be true without proof.

power: Value of an exponent indicating the number of times the base is multiplied by itself.

prime factorization: Process of finding the prime numbers that, when multiplied together, produce a given composite number.

prime number: Whole number larger than 1 that can be divided evenly by only itself and 1.

principal square root: Positive number that when multiplied by itself produces a given positive number. For example, the square roots of 25 are 5 and –5, but the principle square root of 25 is only 5.

prism: A three-dimensional figure that has two parallel, congruent bases; all of its faces are polygons.

product: Result of multiplication.

proper fraction: Fraction whose value is less than one. The numerator is always smaller than the denominator.

proportion: Equation showing that two ratios are equal to one another, such as 1 is to 2 as 3 is to 6 (1:2 = 3:6).

pyramid: A point-topped, three-dimensional figure that has a polygon for its base and triangles for its lateral faces.

Pythagorean theorem: Formula specific to right triangles stating that the hypotenuse *(c)* squared is equal to the sum of the squares of the remaining sides *(a* and *b)*.

quadrant: One of four regions in a coordinate plane defined by the *x*-axis and the *y*-axis.

quadratic: Also known as *second degree*; expression or equation in which the highest power is 2. The degree is 2.

quadrilateral: A four-sided polygon.

quotient: Result of division.

radius: Distance from the center of a circle to its outer edge; half a circle's diameter.

rational number: Quantity, positive or negative, that can be written as a fraction; its decimal equivalent terminates or repeats.

ray: A portion of a line that originates at a point and goes on forever in only one direction.

real number: Any rational or irrational number.

reciprocal: Multiplicative inverse of a given number. The product of a given number and its reciprocal is always 1.

rectangle: Four-sided plane figure with all right angles; its opposite sides are equal to each other in length.

reduce: Process in which a common factor of the numerator and denominator of a fraction is divided out, leaving an equivalent fraction.

regular polygon: A polygon that's both equilateral (equal sides) and equiangular (equal angles).

relatively prime: Two numbers that have no factors in common other than the number 1.

remainder: Value that is left over when one number is divided by another.

resistance: The ratio of voltage to current for an element in an electrical circuit.

resultant: A vector sum.

rhombus: A quadrilateral with four congruent sides.

right angle: 90-degree angle.

right triangle: Three-sided plane figure with a 90-degree or right angle.

root: Value that multiplied by itself a number of times results in the value or number wanted. For example, 2 is the root of 4 because 2 multiplied by itself produces 4.

rounding: Approximating value to the nearest digit or decimal place, such as rounding 14.9 up to 15.

scalar: A constant, or constant multiple.

scalene triangle: A triangle that has no congruent sides.

scientific notation: A standard way or writing very large and very small numbers as the product of two values, or a number between 1 and 10 and a power of 10.

secant: A line that intersects a circle at two points.

sector: A region of a circle bounded by two radii and the arc between those radii; a slice.

similar: Having the same shape; similar figures may be congruent, but they're usually different sizes.

simplify: To combine all that can be combined and put an expression in its most easily understandable form.

slope: Number indicating the measure of a line's steepness or slant and whether it rises or falls.

solution of equation: Value(s) of the variable which make the equation a true statement.

solve: Find the answer or what number the variable stands for.

sphere: A three-dimensional shape in which every point on its surface is equidistant from its center.

square: Four-sided figure in a plane, in which all the sides are the same length and meet at right angles; value with an exponent of 2; perfect square, or product of another number times itself.

square root: Value resulting in a given value when multiplied by itself. For example, the square root of 4 is 2 because $2 \times 2 = 4$, which is a perfect square.

standard volume: Defined as 22.4 liters.

straight angle: A 180-degree angle; basically a line with a point on it.

substitution: Method of replacing a value with its equivalent.

sum: Result of addition.

supplementary angles: Two angles whose measures add up to 180 degrees.

tangent: A line that touches a circle at a single point.

term: Constants, numbers, and variables connected to one another by multiplication or division.

theorem: A mathematical statement that can be proved.

transformation: A passage from one expression, figure, or format to another through a defined process or rule.

transversal: A line that crosses two parallel lines.

trapezoid: A quadrilateral with exactly one pair of parallel sides.

triangle: A three-sided polygon.

trinomial: Expression with three terms. Each term is separated from the others by addition or subtraction.

trisect: To divide a segment or an angle into three congruent parts.

trivial solution: When the solution of a homogenous system has each variable equal to zero.

value: A numeric equivalent or worth of an expression or variable.

variable: Letter representing an unknown number, or what you're solving for in an algebra problem.

vector: A mathematical construct that has both a magnitude and a direction.

velocity: The rate of change in an object's position, expressed as a vector whose magnitude is speed.

vertex: Corner of a figure; where two sides intersect to form an angle.

volume: Measurement of the amount of space within a three-dimensional solid figure.

whole numbers: All natural numbers plus zero. Start with 0 and add 1 repeatedly to find the whole number spectrum (0, 1, 2, 3 . . .).

Index

• *E* •

• **N** •

Business/Accounting Bookkeeping

Bookkeeping For Dummies
978-0-7645-9848-7

eBay Business
All-in-One For Dummies,
2nd Edition
978-0-470-38536-4

Job Interviews
For Dummies,
3rd Edition
978-0-470-17748-8

Resumes For Dummies,
5th Edition
978-0-470-08037-5

Stock Investing
For Dummies,
3rd Edition
978-0-470-40114-9

Successful Time
Management
For Dummies
978-0-470-29034-7

Computer Hardware

BlackBerry For Dummies,
3rd Edition
978-0-470-45762-7

Computers For Seniors
For Dummies
978-0-470-24055-7

iPhone For Dummies,
2nd Edition
978-0-470-42342-4

Laptops For Dummies,
3rd Edition
978-0-470-27759-1

Macs For Dummies,
10th Edition
978-0-470-27817-8

Cooking & Entertaining

Cooking Basics
For Dummies,
3rd Edition
978-0-7645-7206-7

Wine For Dummies,
4th Edition
978-0-470-04579-4

Diet & Nutrition

Dieting For Dummies,
2nd Edition
978-0-7645-4149-0

Nutrition For Dummies,
4th Edition
978-0-471-79868-2

Weight Training
For Dummies,
3rd Edition
978-0-471-76845-6

Digital Photography

Digital Photography
For Dummies,
6th Edition
978-0-470-25074-7

Photoshop Elements 7
For Dummies
978-0-470-39700-8

Gardening

Gardening Basics
For Dummies
978-0-470-03749-2

Organic Gardening
For Dummies,
2nd Edition
978-0-470-43067-5

Green/Sustainable

Green Building
& Remodeling
For Dummies
978-0-470-17559-0

Green Cleaning
For Dummies
978-0-470-39106-8

Green IT For Dummies
978-0-470-38688-0

Health

Diabetes For Dummies,
3rd Edition
978-0-470-27086-8

Food Allergies
For Dummies
978-0-470-09584-3

Living Gluten-Free
For Dummies
978-0-471-77383-2

Hobbies/General

Chess For Dummies,
2nd Edition
978-0-7645-8404-6

Drawing For Dummies
978-0-7645-5476-6

Knitting For Dummies,
2nd Edition
978-0-470-28747-7

Organizing For Dummies
978-0-7645-5300-4

SuDoku For Dummies
978-0-470-01892-7

Home Improvement

Energy Efficient Homes
For Dummies
978-0-470-37602-7

Home Theater
For Dummies,
3rd Edition
978-0-470-41189-6

Living the Country Lifestyle
All-in-One For Dummies
978-0-470-43061-3

Solar Power Your Home
For Dummies
978-0-470-17569-9